List of titles

Already published

Cell Differentiation	J.M. Ashworth
Biochemical Genetics	R.A. Woods
Functions of Biological Membranes	M. Davies
Cellular Development	D. Garrod
Brain Biochemistry	H.S. Bachelard
Immunochemistry	M.W. Steward
The Selectivity of Drugs	A. Albert
Biomechanics	R. McN. Alexander
Molecular Virology	T.H. Pennington, D.A. Ritchie
Hormone Action	A. Malkinson
Cellular Recognition	M.F. Greaves
Cytogenetics of Man and other Animals	A. McDermott
RNA Biosynthesis	R.H. Burdon
Protein Biosynthesis	A.E. Smith
Biological Energy Conservation	C. Jones
Control of Enzyme Activity	P. Cohen
Metabolic Regulation	R. Denton, C.I. Pogson
Plant Cytogenetics	D.M. Moore
Population Genetics	L.M. Cook
Insect Biochemistry	H.H. Rees
A Biochemical Approach to Nutrition	R.A. Freedland, S. Briggs

In preparation

The Cell cycle	S. Shall
Polysaccharides	D.A. Rees
Microbial Metabolism	H. Dalton, R.R. Eady
Bacterial Taxonomy	D. Jones
Molecular Evolution	W. Fitch
Metal Ions in Biology	P.M. Harrison, R. Hoare
Cellular Immunology	D. Katz
Muscle	R.M. Simmons
Xenobiotics	D.V. Parke
Human Genetics	J.H. Edwards
Biochemical Systematics	J.B. Harborne
Biochemical Pharmacology	B.A. Callingham
Biological Oscillations	A. Robertson
Membrane Assembly	J. Haslam
Enzyme Kinetics	P.C. Engel
Functional Aspects of Neurochemistry	G. Ansell, S. Spanner

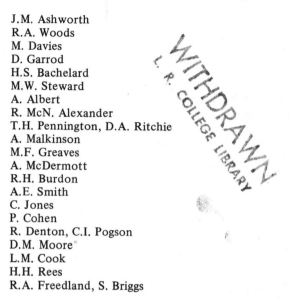

OUTLINE STUDIES IN BIOLOGY

Editor's Foreword

The student of biological science in his final years as an undergraduate and his first years as a graduate is expected to gain some familiarity with current research at the frontiers of his discipline. New research work is published in a perplexing diversity of publications and is inevitably concerned with the minutiae of the subject. The sheer number of research journals and papers also causes confusion and difficulties of assimilation. Review articles usually presuppose a background knowledge of the field and are inevitably rather restricted in scope. There is thus a need for short but authoritative introductions to those areas of modern biological research which are either not dealt with in standard introductory textbooks or are not dealt with in sufficient detail to enable the student to go on from them to read scholarly reviews with profit. This series of books is designed to satisfy this need. The authors have been asked to produce a brief outline of their subject assuming that their readers will have read and remembered much of a standard introductory textbook of biology. This outline then sets out to provide by building on this basis, the conceptual framework within which modern research work is progressing and aims to give the reader an indication of the problems, both conceptual and practical, which must be overcome if progress is to be maintained. We hope that students will go on to read the more detailed reviews and articles to which reference is made with a greater insight and understanding of how they fit into the overall scheme of modern research effort and may thus be helped to choose where to make their own contribution to this effort. These books are guidebooks, not textbooks. Modern research pays scant regard for the academic divisions into which biological teaching and introductory textbooks must, to a certain extent, be divided. We have thus concentrated in this series on providing guides to those areas which fall between, or which involve, several different academic disciplines. It is here that the gap between the textbook and the research paper is widest and where the need for guidance is greatest. In so doing we hope to have extended or supplemented but not supplanted main texts, and to have given students assistance in seeing how modern biological research is progressing, while at the same time providing a foundation for self help in the achievement of successful examination results.

J.M. Ashworth, Professor of Biology, University of Essex.

Plant Cytogenetics

D.M. Moore

Professor of Botany
University of Reading

LONDON
CHAPMAN AND HALL

A Halsted Press Book

JOHN WILEY & SONS, INC., NEW YORK

Preface

One of the great unifying themes of biology during the past seventy years or so has been provided by genetics and cytology, which have together furnished a firm basis for understanding the materials and processes upon which the variation and evolution of all living organisms depends. Many recent texts have, quite rightly, stressed those tenets of cytogenetics common to most eukaryotes and it might be wondered, therefore, why animals and plants have been considered in separate volumes of this series. There are two principal reasons for this. Firstly, there is now so much cytogenetical information on plants and animals that any attempt to even outline the subject in a single slim volume would result in an unacceptably superficial treatment. Secondly, whilst acknowledging the common cytogenetical foundations of all organisms, stemming in part from the virtually ubiquitous genetic material, DNA, it is sometimes forgotten that plants and animals are different, with different evolutionary opportunities and, in many instances, they have utilized their common genetical and chromosomal endowment to adopt different evolutionary strategies, which are reflected in their patterns of variation. Consequently, some cytogenetical processes are better observed in animals, others in plants. In this book I have attempted to outline some of the features of the structure and behaviour of chromosomes, which still encompass the central enigmas of cytogenetics, and have then continued by looking at various chromosmal mechanisms found in plants and their role in generating and canalizing the variation which, as a student of taxonomy and evolution, is the real reason for my interest in cytogenetics. Of necessity, such an outline as this is selective, even when concerned almost entirely with flowering plants as in this case, but I hope that the references supplied will encourage the reader to consult the original sources which span the history of cytogenetics and which, above all, permit an appreciation of the large amount of work that has already been carried out and of the great task still ahead.

I should like to thank Professor V.H. Heywood for his extremely helpful comments on the completed text. I am also very grateful to Mrs Abigail Gillett and Mrs Rosa Husain for skilfully transcribing my handwriting into an orderly typescript. Finally, thanks are due to my wife and children for tolerating me while I wrote it.

D.M.M.

Plant Cytogenetics

D.M. Moore

Professor of Botany
University of Reading

LONDON
CHAPMAN AND HALL

A Halsted Press Book
JOHN WILEY & SONS, INC., NEW YORK

First published in 1976
by Chapman and Hall Ltd.
11 New Fetter Lane, London EC4P 4EE
© *1976 D.M. Moore*
Printed in Great Britain at the
University Printing House, Cambridge

ISBN 0 412 13440 3

Distributed in the U.S.A.
by Halsted Press, a Division
of John Wiley & Sons, Inc. New York

Library of Congress Cataloging in Publication Data

Moore, David Moresby
    Plant cytogenetics.

    (Outline studies in biology)
    'A Halsted Press book.'
    1. Plant cytogenetics. I. Title.

QH433.M66   1976   581.1'51   76-15967
ISBN 0-470-15146-3

Contents

Preface

One of the great unifying themes of biology during the past seventy years or so has been provided by genetics and cytology, which have together furnished a firm basis for understanding the materials and processes upon which the variation and evolution of all living organisms depends. Many recent texts have, quite rightly, stressed those tenets of cytogenetics common to most eukaryotes and it might be wondered, therefore, why animals and plants have been considered in separate volumes of this series. There are two principal reasons for this. Firstly, there is now so much cytogenetical information on plants and animals that any attempt to even outline the subject in a single slim volume would result in an unacceptably superficial treatment. Secondly, whilst acknowledging the common cytogenetical foundations of all organisms, stemming in part from the virtually ubiquitous genetic material, DNA, it is sometimes forgotten that plants and animals are different, with different evolutionary opportunities and, in many instances, they have utilized their common genetical and chromosomal endowment to adopt different evolutionary strategies, which are reflected in their patterns of variation. Consequently, some cytogenetical processes are better observed in animals, others in plants. In this book I have attempted to outline some of the features of the structure and behaviour of chromosomes, which still encompass the central enigmas of cytogenetics, and have then continued by looking at various chromosmal mechanisms found in plants and their role in generating and canalizing the variation which, as a student of taxonomy and evolution, is the real reason for my interest in cytogenetics. Of necessity, such an outline as this is selective, even when concerned almost entirely with flowering plants as in this case, but I hope that the references supplied will encourage the reader to consult the original sources which span the history of cytogenetics and which, above all, permit an appreciation of the large amount of work that has already been carried out and of the great task still ahead.

I should like to thank Professor V.H. Heywood for his extremely helpful comments on the completed text. I am also very grateful to Mrs Abigail Gillett and Mrs Rosa Husain for skilfully transcribing my handwriting into an orderly typescript. Finally, thanks are due to my wife and children for tolerating me while I wrote it.

D.M.M.

1 The beginnings of cytogenetics

1.1 Rise of the chromosomes

Cytology, the scientific study of cells, had its beginnings in the 17th century, when the first microscopes were used by Hooke (1635–1703), Grew (1641–1712) and Malpighi (1628–1694) to make the initial observations which eventually led to the theory of Schleiden and Schwann (1838–1839) that the cell was the basic unit of structure and function in all living organisms. The general introduction of compound microscopes about this time permitted rapid progress in cytology so that by 1858 Remak and Virchow could suggest that all cells arose from the division of pre-existing cells. Increasing appreciation of the importance of the nucleus led, with the observations of Hertwig [1[(1875) on sea urchin eggs, to the recognition of its role in fertilization and cell-division.

During the next few years the role of the chromosomes in the nuclear cycle was realized and to some extent described. In both plants and animals Flemming, van Beneden and Strasburger observed and described mitosis (Fig. 1.1), as well as the salient features of meiosis [6,7], which was more fully described by von Winiwarter [2] (Fig. 1.2), while Balbiani [3] and Carnoy [4] discovered and observed the salivary gland chromosomes of Diptera. Van Beneden showed that during mitosis the daughter halves of the chromosomes pass to opposite poles and that the fertilized egg of *Ascaris* receives an equal number of chromosomes from each parent, a number halved during the meiotic divisions which precede the formation of the gametes so that it remains constant from one generation to the next. Although cytology still remained a branch of either histology or embryology during this period there was, largely because of the influence of Roux [5] and Weismann [6], a gradual acceptance of the idea that the chromosomes were the material basis of heredity. This, then, was the state of knowledge when E.B. Wilson [7] wrote the second edition of his great work, *The Cell in Development and Inheritance,* published in the year that Mendel's genetic discoveries were disinterred and made available to the scientific community.

1.2 Appearance of genetics

Although it is usual to trace the history of genetics back to Aristotle, and even Hippocrates, who recognized that individuals may resemble remote ancestors rather than their parents and that the effects of mutilations are not transmitted to offspring, the foundations of the subject which persisted into modern times were laid during the 18th century. Thus, Kölreuter, who published the results of his extensive crosses on plants between 1761 and 1766, recognized that hybrids were usually intermediate between the parents and that they were often sterile in crosses between widely different forms; he also emphasized the identity of hybrids from reciprocal crosses, while even earlier Robert Fairchild (1719) observed the dominance of double over single flowers in the progeny of crosses in *Dianthus* [8]. The continued accumulation of data on animals and plants derived from gardeners, farmers and sportsmen, during the next hundred years was brought together by Darwin in *The Variation in Animals and Plants under Domestication* (1868), an interesting source of information in which he, like Gaertner (1772–1850) emphasized the greater variability of the second and later generations compared to the first generation resulting from hybridization. Most

2.1.2 Ultrastructure

Although, as noted above, the problem of how the DNA molecule, with its fundamental genetical properties, is integrated into the architecture of the chromosome is still incompletely resolved, histochemical and electron microscope studies are gradually clarifying the picture. The study of electron micrographs of mitotic and meiotic chromosomes from various plants and animals led to the conclusion that the chromosome is built like a cable with numerous identical strands [7,8]. Subsequent work, which has employed both electron microscopic study of sections and of nuclei (spread on an air-water interface, picked up on carbon-coated grids, fixed and dried by, for example, amylacetate), has confirmed that the chromosome is composed of fibrils [9]. The reported diameters of these have varied from 3–50 nm, the most frequent range being 10–20 nm, but their length cannot be determined. However, Ris and others have shown that the diameter of the fibrils varies considerably, depending upon the use of different buffers during fixation [10], and there is now considerable evidence from electron microscopy and X-ray diffraction data that fibre diameters are about 10 nm [11,12].

Although a detailed consideration of the relationship between DNA and histones in the chromosomes of plants and other eukaryotes is beyond the scope of this book, and is indeed as yet unresolved, it is worth pointing out that X-ray and chemical data [12,13] suggest that chromatin fibrils are composed of a series of repeating units consisting of tightly packed DNA and associated protein, alternating with more extended DNA and associated protein. Electron micrographs of chromatin fibrils following formaldehyde fixation show a bead-like appearance. The thickened 'beads' are about 7 nm in diameter, which is compatible with a globular histone tetramer (diameter 4–5 nm) associated with a double helix of DNA (diameter *c.* 2 nm, [12,14]). Cleavage of chromatin by certain nucleases produces pieces of DNA comprising about 200 base pairs, or multiples thereof, and it is suggested that this is consistent with biochemical and X-ray data on the size of the repeating units. There is, then, some evidence that the chromatin fibril is a flexibly jointed chain of repeating units; this flexibility would permit the extensive coiling and folding of which the chromatin fibril is known to be capable.

The relationship between the apparent multiple fibrillar structure of the chromonemata and the classical evidence of the chromatid as the basic unit of cytogenetics has long proved difficult to reconcile. Although some workers [15] have considered the chromatids to consist of a single, strongly folded and coiled chromatin fibril there is considerable evidence to support the view that the chromatid contains at least 2 DNA duplexes [16]. Half-chromatids have been reported from light microscopic studies in *Endymion, Haemanthus* and *Vicia* while following irradiation there is cytogenetical evidence of subchromatid breaks and recombination (Section 3.2, 3.3). Reconciliation between these two apparently conflicting situations may depend upon the observation that, in studies of the synaptinemal complex during pachytene (Section 3.3), only a part of the chromatin is involved in recombination. Perhaps, therefore, the multiple structure is a form of genetic insurance policy, but much cytogenetical and cytochemical information is still needed to fully understand the ultrastructure of the chromosome.

2.2 Centromeres, telomeres and chromosome form

2.2.1 Centromeres

The centromere (kinetochore) is the most conspicuous feature of most chromosomes, appearing from mitotic prometaphase to anaphase as a region which, because it does not coil, is a weakly stained 'primary constriction' distinguished from the thicker, darker staining chromosome arms. The distinctness of the centromere varies a great deal between different organisms but it can generally be enhanced by

1 The beginnings of cytogenetics

1.1 Rise of the chromosomes

Cytology, the scientific study of cells, had its beginnings in the 17th century, when the first microscopes were used by Hooke (1635–1703), Grew (1641–1712) and Malpighi (1628–1694) to make the initial observations which eventually led to the theory of Schleiden and Schwann (1838–1839) that the cell was the basic unit of structure and function in all living organisms. The general introduction of compound microscopes about this time permitted rapid progress in cytology so that by 1858 Remak and Virchow could suggest that all cells arose from the division of pre-existing cells. Increasing appreciation of the importance of the nucleus led, with the observations of Hertwig [1 [(1875) on sea urchin eggs, to the recognition of its role in fertilization and cell-division.

During the next few years the role of the chromosomes in the nuclear cycle was realized and to some extent described. In both plants and animals Flemming, van Beneden and Strasburger observed and described mitosis (Fig. 1.1), as well as the salient features of meiosis [6,7], which was more fully described by von Winiwarter [2] (Fig. 1.2), while Balbiani [3] and Carnoy [4] discovered and observed the salivary gland chromosomes of Diptera. Van Beneden showed that during mitosis the daughter halves of the chromosomes pass to opposite poles and that the fertilized egg of *Ascaris* receives an equal number of chromosomes from each parent, a number halved during the meiotic divisions which precede the formation of the gametes so that it remains constant from one generation to the next. Although cytology still remained a branch of either histology or embryology during this period there was, largely because of the influence of Roux [5] and Weismann [6], a gradual acceptance of the idea that the chromosomes were the material basis of heredity. This, then, was the state of knowledge when E.B. Wilson [7] wrote the second edition of his great work, *The Cell in Development and Inheritance,* published in the year that Mendel's genetic discoveries were disinterred and made available to the scientific community.

1.2 Appearance of genetics

Although it is usual to trace the history of genetics back to Aristotle, and even Hippocrates, who recognized that individuals may resemble remote ancestors rather than their parents and that the effects of mutilations are not transmitted to offspring, the foundations of the subject which persisted into modern times were laid during the 18th century. Thus, Kölreuter, who published the results of his extensive crosses on plants between 1761 and 1766, recognized that hybrids were usually intermediate between the parents and that they were often sterile in crosses between widely different forms; he also emphasized the identity of hybrids from reciprocal crosses, while even earlier Robert Fairchild (1719) observed the dominance of double over single flowers in the progeny of crosses in *Dianthus* [8]. The continued accumulation of data on animals and plants derived from gardeners, farmers and sportsmen, during the next hundred years was brought together by Darwin in *The Variation in Animals and Plants under Domestication* (1868), an interesting source of information in which he, like Gaertner (1772–1850) emphasized the greater variability of the second and later generations compared to the first generation resulting from hybridization. Most work when it was rediscovered and the results confirmed by Correns, de Vries and von Tchermak [11] in 1900.

1.3 Chromosomal theory of inheritance

Although the relationship between chromosomes and genes was suspected at the time Mendel's work was rediscovered, it took a further three years to establish the interpretation which has persisted to the present. During that period Montgomery (1901) and Sutton (1902), working on grasshoppers, showed that chromosomes occur in distinct pairs, often of recognizable shape and size, and that synapsis involves the union of maternal and paternal chromosomes, while Winiwarter (1901) concluded, from his studies of meiosis in rabbit ovaries, that bivalents in the first meiotic division resulted from the chromosomes pairing side-by-side and not end to end as believed by Weismann and others [2]. Boveri (1902) showed, from his studies of polyspermy in the fertilization of sea-urchin eggs, that the chromosomes of an individual were not equivalent to one another and that a full complement is necessary for normal development of the cell. Correns and Cannon, both in 1902, pointed out the close parallelism between Mendelian segregation and chromosome reduction, concluding that the genes are on the chromosomes; but they, like de Vries a year later, were incorrect in many suppositions, such as their view that maternal and paternal chromosomes went to opposite poles during meiosis [11]. In the same years two papers by Guyer showed an understanding that random assortment between different pairs of chromosomes would give the independent assortment of genes required by Mendel, although the cytological demonstration was not made until 1913 (Carothers [12], Fig. 1.3). It was, however, Sutton who, in 1903 brought together the data from cytology and genetics to clearly show the role of the chromosomes in heredity and hence to firmly establish the field of cytogenetics. Boveri, in a paper published the same year, advanced many of the same ideas so that the hypothesis

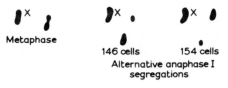

Metaphase

146 cells   154 cells
Alternative anaphase I
segregations

Fig. 1.3 Independent segregation of unpaired X chromosome and heteromorphic pair of large and small chromosomes observed by Carothers [12] at meiotic anaphase in *Brachystola*.

correlating gene and chromosome transmission is known as the 'Sutton-Boveri Hypothesis'.

Basically, the hypothesis is as follows:—

1. In somatic cells there are two similar groups of chromosomes, one of maternal and one of paternal origin. This occurrence of chromosomes in homologous pairs parallels the occurrence of genes in pairs.

2. The chromosomes retain a morphological individuality throughout the various cell-divisions; genes show a similar continuity.

3. During meiosis homologous pairs of chromosomes are brought together and then the members of each pair segregate into different germ cells independently of the members of other pairs; Mendelian genes segregate independently at some time prior to gamete formation.

4. Each chromosome, or chromosome-pair, has a definite role in the life and development of the individual.

In addition to establishing the relationship between genes and chromosomes, Sutton recognized that there must be non-independent assortment of some genes (*linkage*) otherwise, as he noted, 'the numbers of distinct characters ... could not exceed the number of chromosomes'.

The association of a particular inherited character with a particular chromosome was made between 1901 and 1906 by McClung, Stevens, Wilson and others who showed that in Hemiptera and Orthoptera, females have one more chromosome than the males [3]. This so-called X-chromosome occurs in all eggs but in only 50% of sperm so that half of the resultant zygotes are XX and female, while half are X0 and male. The presence of a small Y-chromosome, partially homologous with the X-chromosome, in males of beetles, insects, mammals and other groups confirmed the same pattern — that the sex chromosomes of the male gametes determine the sex of the progeny — while the discovery of the reverse situation, female heterozygosity, in birds and lepidoptera confirmed the importance of chromosomes in sex determination.

The association of a particular gene with a particular chromosome was demonstrated by Morgan [14], who showed that the inheritance of the recessive allele (w) for white-eye in *Drosophila* paralleled that of the X-chromosome (Fig. 1.4). Conclusive evidence that the white locus was situated on the X-chromosome was provided by Bridges [15], who found that sometimes a cross between a white-eyed female and a red-eyed male gave an occasional white-eyed female or red-eyed male among the F$_1$ progeny. This was found to be due to non-separation of the X-chromosomes at meiosis in the female so that, exceptionally, eggs with either two or no X-chromosomes were produced. The consequences of this are shown in Fig. 1.5, the XXY constitution of the white-eyed females being confirmed cytologically. With this and other genes present on the X-chromosomes, Bridges found the correlation between genetic-

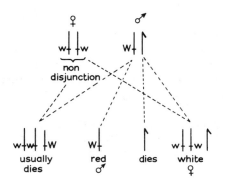

Fig. 1.5 Results of cross between white-eye female (*ww*), with non-disjunction of X-chromosomes, and red-eye male *Drosophila*.

al and chromosomal inheritance to be exact, thus providing the first critical evidence that genes are on chromosomes.

1.4 Linkage, crossing-over and chromosome maps

As noted above, Sutton pointed out that if there were more gene loci than chromosomes, a fact since abundantly demonstrated in all plants and animals studied at all intensively, then his theory would not permit the Mendelian law of independent segregation to apply to genes located on the same chromosome. The data of Bateson and Punnet [16] on Sweet Peas provided genetical evidence of this linkage,

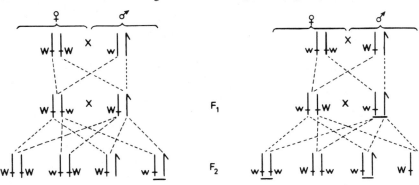

Fig. 1.4 Inheritance of alleles for white (*w*) and red (*W*) eyes in *Drosophila*. Reciprocal crosses show parallel transmission of *W* allele and X-chromosome (⊤). Y-chromosome indicated by ⊦⎰. White eye phenotypes underlined.

11

while Sturtevant [17] demonstrated the linear arrangement of genes on the chromosome and initiated the use of the 3-point testcross for mapping the loci, both soon brought to cytological reality by Painter's [18] manipulation of salivary-gland chromosomes and Muller's [19] discovery that X-rays can simultaneously mutate genes and alter chromosomal structure. Finally, Creighton and McClintock [20] demonstrated the correlation between genetical recombination and cytological crossing-over (Fig. 1.6) in maize and so brought into prominence a cytogenetical mechanism which is still not fully understood more than 40 years later.

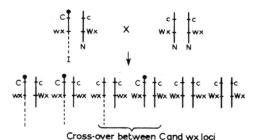

Cross-over between C and wx loci

Fig. 1.6 Cross in maize demonstrating correlation between recombination and crossing over. One plant is heterozygous for a chromosome (I) with a terminal knob and a long extra segment (- - - -) and a normal chromosome (N), and for alleles determining waxy (*wx*) or starchy (*Wx*) and coloured (*C*) or colourless (*c*) endosperm.

References

[1] Hertwig, O. (1875), *Morph. Jahrb.*, **1**, 347-434.
[2] Whitehouse, H.L.K. (1965), *The Mechanism of Heredity*, Chap. 6, Arnold, London.
[3] Balbiani, E.G. (1881), *Zool. Anz.*, **4**, 637-641, 662-666.
[4] Carnoy, J.B. (1884), *La Biologie Cellulaire*,
[5] Roux, W. (1883), *Über die Bedeutung der Kerntheilungsfiguren*, Engelmann, Leipzig.
[6] Weismann, A. (1889), *Essays upon Heredity*, Oxford University Press, Oxford.
[7] Wilson, E.B. (1900), *The Cell in Development and Heredity*, edit. 2., Macmillan, New York.
[8] Zirkle, C. (1951), In : *Genetics in the 20th Century*, Dunn, L.C. (ed), Macmillan, New York
[9] Johannsen, W. (1909), *Elemente der exakten Erblichkeitslehre*, Fischer, Jena.
[10] Bateson, W. and Saunders, E.R. (1902), *Rep. Evol. Cttee Roy. Soc.*, **2**, 1-55.
[11] Sturtevant, A.H. (1965), *A History of Genetics*, Harper and Row, New York.
[12] Carothers, E.E. (1913), *J. Morph.*, **24**, 487-511.
[13] Sutton, W.S. (1903), *Biol. Bull.*, **4**, 231-251.
[14] Morgan, T.H. (1910), *Science*, **32**, 120-122.
[15] Bridges, C.B. (1916), *Genetics*, **1**, 1-52, 107-163.
[16] Bateson, W. and Punnett, R.C. (1905-08), *Rep. Evol. Cttee Roy. Soc.*, **2-4**.
[17] Sturtevant, A.H. (1916), *J. Expl. Zool.*, **14**, 41-59.
[18] Painter, T.S. (1933), *Science*, **78**, 585-586.
[19] Muller, H.J. (1927), *Science*, **66**, 84-87.
[20] Creighton, H.B. and McClintock, B. (1931), *Proc. Nat. Acad. Sci. U.S.A.* 17, 492-497.

2 Chromosome structure

Although deoxyribosenucleic acid (DNA) had been shown to be largely localized in the chromosomes (it also occurs in mitochondria and other cell organelles) by the specific staining techniques of Feulgen and Rossenbeck [1] it was a further 20 years before Avery, Macleod and McCarty demonstrated it to be the primary hereditary material. When, in 1953, Watson and Crick proposed their double helix, which would permit accurate pairing and duplication of DNA, as well as suggesting how mutation might occur, the central problems in cytogenetics seemed to be solved [2]; indeed, studies of the 'chromosome' (*genophore*) [3] in bacteria and other prokaryotes give credence to this. However, the chromosomes of most animals and plants (*eukaryotes*) each contain much more DNA, arranged in a linear and not a circular fashion. Furthermore, unlike genophores, the DNA in chromosomes is regularly and intimately associated with histone molecules, although the manner of their arrangement is still largely unresolved and constitutes one of the major problems of cytogenetics. Information on chromosome structure is derived from four sources:— (a) light microscopy, using bright field or phase contrast illumination; (b) electron microscopy; (c) cytochemistry; (d) genetic behaviour.

2.1 Chromonemata and chromatids

2.1.1 Gross structure
The most generally accepted basic units of chromosomal organization appear in the interphase nucleus as a series of fine threads visible only by interference microscopy. Once the cell commences division these chromonemata shorten and increase in volumes as they coil so that they become visible by phase-contrast microscopy or, following fixation and staining with fuchsin, or carmine, or orcein etc., with the ordinary light microscope. They are shown to be associated in pairs, attached to a single centromere, to form the chromosome; during this visible phase they are known as chromatids. Exceptionally, the chromosome may be polytene and consist of many chromatids laterally opposed, as in the salivary glands of *Drosophila* and other insects, and also some plant cells (Section 3.5).

During meiosis the chromonemata exhibit a chromomeric pattern, which is not seen in mitotic division. The chromomeres appear at first prophase as a series of darker staining 'beads' on the chromonemata. They may be of uniform size, show a regular gradation with larger knobs near the centromere and smaller spots towards the chromosome ends, as in tomato, or form a less distinct size-gradient, as in rye and *Salvia viridis* [4,5]. The basic pattern of chromomeres is characteristic for each chromosome. La Cour and Wells [6] have shown from light and electron microscope studies of leptotene chromosomes in *Tulbaghia*, *Fritillaria* and *Lilium*, that the chromomeres are borne eccentrically to the chromosome axis. As meiotic prophase proceeds the chromomeres increase in size and decrease in number as they tend to merge into one another, the number diminishing in proportion to the chromosome length. This is explained by the generally accepted view that the chromomeres are coiled portions of the chromonemata; in *Tradescantia*, for example, the chromomeres increase in size until they become the visibly distinct coils of late prophase.

2.1.2 Ultrastructure

Although, as noted above, the problem of how the DNA molecule, with its fundamental genetical properties, is integrated into the architecture of the chromosome is still incompletely resolved, histochemical and electron microscope studies are gradually clarifying the picture. The study of electron micrographs of mitotic and meiotic chromosomes from various plants and animals led to the conclusion that the chromosome is built like a cable with numerous identical strands [7,8]. Subsequent work, which has employed both electron microscopic study of sections and of nuclei (spread on an air-water interface, picked up on carbon-coated grids, fixed and dried by, for example, amylacetate), has confirmed that the chromosome is composed of fibrils [9]. The reported diameters of these have varied from 3–50 nm, the most frequent range being 10–20 nm, but their length cannot be determined. However, Ris and others have shown that the diameter of the fibrils varies considerably, depending upon the use of different buffers during fixation [10], and there is now considerable evidence from electron microscopy and X-ray diffraction data that fibre diameters are about 10 nm [11,12].

Although a detailed consideration of the relationship between DNA and histones in the chromosomes of plants and other eukaryotes is beyond the scope of this book, and is indeed as yet unresolved, it is worth pointing out that X-ray and chemical data [12,13] suggest that chromatin fibrils are composed of a series of repeating units consisting of tightly packed DNA and associated protein, alternating with more extended DNA and associated protein. Electron micrographs of chromatin fibrils following formaldehyde fixation show a bead-like appearance. The thickened 'beads' are about 7 nm in diameter, which is compatible with a globular histone tetramer (diameter 4–5 nm) associated with a double helix of DNA (diameter $c.$ 2 nm, [12,14]). Cleavage of chromatin by certain nucleases produces pieces of DNA comprising about 200 base pairs, or multiples thereof, and it is suggested that this is consistent with biochemical and X-ray data on the size of the repeating units. There is, then, some evidence that the chromatin fibril is a flexibly jointed chain of repeating units; this flexibility would permit the extensive coiling and folding of which the chromatin fibril is known to be capable.

The relationship between the apparent multiple fibrillar structure of the chromonemata and the classical evidence of the chromatid as the basic unit of cytogenetics has long proved difficult to reconcile. Although some workers [15] have considered the chromatids to consist of a single, strongly folded and coiled chromatin fibril there is considerable evidence to support the view that the chromatid contains at least 2 DNA duplexes [16]. Half-chromatids have been reported from light microscopic studies in *Endymion, Haemanthus* and *Vicia* while following irradiation there is cytogenetical evidence of subchromatid breaks and recombination (Section 3.2, 3.3). Reconciliation between these two apparently conflicting situations may depend upon the observation that, in studies of the synaptinemal complex during pachytene (Section 3.3), only a part of the chromatin is involved in recombination. Perhaps, therefore, the multiple structure is a form of genetic insurance policy, but much cytogenetical and cytochemical information is still needed to fully understand the ultrastructure of the chromosome.

2.2 Centromeres, telomeres and chromosome form

2.2.1 Centromeres

The centromere (kinetochore) is the most conspicuous feature of most chromosomes, appearing from mitotic prometaphase to anaphase as a region which, because it does not coil, is a weakly stained 'primary constriction' distinguished from the thicker, darker staining chromosome arms. The distinctness of the centromere varies a great deal between different organisms but it can generally be enhanced by

pre-treatment with mitotic inhibitors, such as paradichlorobenzene and colchicine, before staining. Although the position of the centromere is constant for a given chromosome it can vary between them thus providing a valuable marker for describing the chromosome complement. Although more elaborate classifications have been proposed [17], it is customary to distinguish three principal chromosome types based on the position of the centromere:—
metacentric — the median centromere separates two arms of approximately equal length;
acrocentric — the interstitial centromere separates two arms of obviously unequal length;
telocentric — the centromere is terminal to give a one-armed chromosome.

There has been much discussion as to whether the centromere is ever truly terminal but, even if this is so, there are certainly many cases in which one arm is not visible and the chromosomes are apparently telocentric (Fig. 2.1b; Section 5.3).

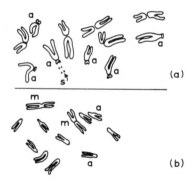

Fig. 2.1 Chromosome types in karyotypes of two plant species. (a) *Callisia fragrans* (2n = 12) −6 metacentrics + 6 acrocentrics (a), one with satellites (s). (Drawn from Jones and Jopling, [11].) (b) *Oxalis dispar* (2n = 12) −2 metacentrics (m) + 2 acrocentrics (a) + 8 telocentrics. (Drawn from Marks, [53].)

Whilst examining cleavage divisions in the salamander, Metzner described differential staining of small bodies within the centromere These 'Leitkörperchen' have subsequently rted in many animals and plants

and variously referred to as polar granules, attachment chromomeres and kinetochores [19,20,21]. There has been a tendency to use the terms centromere and kinetochore synonymously for the primary constriction but nowadays the term kinetochore is reserved for the structures within the centromere by which the chromosomes are moved during cell division (Section 3.3). Lima-de-Faria [22] showed that the centromere contained two to five pairs of darker staining chromomeres joined by uncoiled chromonemal fibrils (Fig. 2.2). This symmetrical organization gives two mirror images about a plane passing

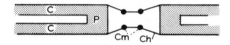

Fig. 2.2 Lima-de-Faria's [22] model of the centromere. Pairs of centromeric chromomeres (Cm) are connected to each other and to the proximal regions (P) joining the sister chromatids (C) by the chromatonemal fibrils (Ch).

through the centre of the centromere and this structure has been reported in many plants, including *Allium cepa* and *Tradescantia* spp. [23]. However, electron microscopy of sectioned material shows that practically all metaphase chromosomes can be interpreted as having one more or less circular kinetochore per chromatid [24,25,26]. In species of *Tradescantia, Ornithogalum, Rhoeo* and *Allium* the kinetochores, when stained by Giemsa techniques (Section 2.3), appear during midprophase to prometaphase. They are about 0.5 μm in diameter and sometimes appear to be attached to remnants of spindle fibres. By metaphase the two kinetochores, one per chromatid, are clearly separated and lie laterally on opposite sides of the centromeres [27]. The kinetochore is believed to be rich in repetitive DNA, which would permit its division, suggested by observations on chromosome structural changes (Chapter 5), but it would seem that no more than two of

Lima-de-Faria's 'centromeric chromomeres' can be kinetochores.

In some organisms the centromere is unlocalized so that many points along the chromosome can function as kinetochores and it has been shown that the diffuse or polycentric chromosomes of, for example, *Luzula* and *Cyperus* do not possess kinetochores of the kind described above [28].

2.2.2 Telomeres

If chromosomes are fractured by X-rays, for example, the resulting segments may fuse again; however, they will not fuse with the ends of chromosome arms, which themselves cannot fuse with each other. This has led to the consideration that the chromosome is terminated by a telomere which confers polarity upon it. The telomere has been shown [29] to be a compound structure consisting of several differentiated segments; in rye, for example, it is composed of at least 2 pairs of chromomeres and intercalary fibrils. Breakage within this region should still give stable chromosome ends, as has been demonstrated in rye and maize. The compound structure of the telomere is very like that of the centromere and they are further shown to share a number of properties relating to their cycle of division and behaviour during meiosis (Sections 3.3, 3.4).

2.2.3 Secondary constrictions and satellites

In many chromosome complements at least one pair of chromosomes is seen to have an unspiralized, non-staining region additional to the centromere. This secondary constriction frequently occurs near the end of the chromosome so that the segment beyond the constriction is small and is then termed a satellite or trabant, joined to the rest of the chromosome by the satellite stalk. The only known function of the secondary constriction is that of nucleolar organization and it is believed that the nucleoli, involved in protein synthesis, are controlled by specific loci associated with the secondary constriction; in maize this nucleolar organizing element is at the base of the satellite stalk and

it has been shown that, following fragmentation by X-irradiation, both sub-units remain functional [30], thus indicating a compound structure as in centromeres and telomeres.

The number of satellited chromosomes in the complement varies in different organisms and is not always parallelled by the number of nucleoli visible at prophase, possibly because of their coalescence. Similarly, chromosomes associated with nucleolar organization may not possess satellites, as in *Nothoscordum inutile,* for example, in which the four chromosomes concerned with this activity seem to have compound constrictions with both centromeric and nucleolar organizing functions, while in several *Trillium* species, the nucleolar organizer appears to be terminal [31].

A further point to remember when using satellites to characterize features of a chromosome complement is that the secondary constriction can vary greatly during the course of mitosis. Thus, during prometaphase the satellites are joined to the chromosome arm by a long slender satellite stalk, which is readily fractured during squash preparations, while at metaphase, particularly following pretreatment with drugs such as oxyquinoline, the stalk can be so short as to make it difficult to determine the presence of the satellite [32]. Further, homologous chromosomes can differ in the size of their satellites. The nucleolar organizer of one species may be dominant to that of another species so that in hybrids the chromosomes of the 'weaker' set may not show the satellite present in the parent.

2.3 Euchromatin and heterochromatin

The standard sequence of condensation (at its maximum during metaphase — anaphase) and elongation (greatest during interphase), described (Figs. 1.1, 1.2) for the chromonemata during the nuclear cycle is known as the eucycle and the chromatin which follows this sequence is termed euchromatin. However, some chromosome segments, and even whole chromosomes, have a different cycle they are more condensed and deep

16

than the remainder of the complement at prophase, for example, (*positively heteropycnotic*) and negatively heteropycnotic later in nuclear division when they appear under-condensed and lightly stained. Such allocyclic chromosomes or segments are described as hererochromatic and their occurrence and position can provide useful markers in charting the karyotype of a population or species, (Fig. 2.3).

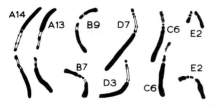

Fig. 2.3 Heterochromatin segments (unshaded) in chromosome complement of *Trillium kamschaticum* after cold treatment. Letters indicate homologous chromosomes, which may be homozygous or heterozygous for banding patterns (indicated by numbers). (After Haga and Kurabayashi, [54].)

In addition to its heteropycnosis, heterochromatin has been shown to differ from euchromatin in its behaviour and, perhaps, structure, since it is inactive in transcription and contains late-replicating, in some cases definitely repetitive, DNA [33,34,35]. Heterochromatin is now generally divided into two classes [36] :— regions of *constitutive heterochromatin* regularly show the features described above and have few or no structural genes, while *facultative heterochromatin* contains structural genes but is cytologically condensed and genetically inactive and occurs only in particular tissues, since it is apparently dependent upon certain physiological and developmental conditions. Nucleic acid hybridization techniques have demonstrated a biochemical difference between these two types of hetero-⸱⸱⸱⸱matin [37]. The conversion of eu- to [18]. ⸱⸱⸱⸱matic regions and vice versa, as a ly been rep⸱⸱⸱tion effect, may refer to facultative

heterochromatin but the situation is not clear.

The visibility of heterochromatic regions varies greatly between organisms and this is possibly related to the amount of heterochromatin in the chromosome or in the whole complement [31]. In *Paris japonica* [38] and *Trillium undulatum* [39], for example, H (heterochromatic)-segments are readily apparent at normal temperatures but in many species of *Fritillaria, Hordeum, Secale, Tulbaghia, Vicia,* etc. it has been found that such segments only become visible after the plants have been subjected to cold treatments of 4–6°C for about 72 hours [40,41], perhaps by a process of localized uncoiling [42]. Such cold treatment has permitted detailed studies of H-banding patterns which have proved of considerable value in characterizing the karyotypes of individuals and populations in relation to taxonomic and evolutionary work.

During recent years various techniques for the preferential staining of heterochromatin have been developed. The most effective stains are Giemsa, first used with mouse chromosomes [43] and subsequently widely employed in human and animal cytogenetics, and various quinacrine dyes which fluoresce when viewed under UV light [44]. With the adaptation of these techniques to plant material [45] staining by Giemsa and various fluorochromes, such as quinacrine, quinacrine mustard, ethidium bromide and bibenzimidazoles, has now been applied to a range of taxa. Although much still remains to be learned, some general points about such preferential staining can now be made.

In *Trillium grandiflorum, Fritillaria recurva, F.lanceolata, Scilla sibirica* and *Vicia faba* the most strongly Giemsa stained chromosome segments were shown to be identical with the heterochromatic regions revealed by cold treatment [46]. Care has to be taken with the pretreatment with heat and alkali, since the contrast between eu- and heterochromatin disappeared in *Scilla sibirica* under certain conditions or if staining was prolonged. In *Vicia faba* different pretreatments produced different banding patterns and no routine technique would allow simultaneous appearance of all the bands within the karyotype. This suggested that different classes of heterochromatin occur in the H-segments, and work on quinacrine fluorescence of *Scilla sibirica* chromosomes supports this. Bands located near the centromere showed enhanced fluorescence, while bands with reduced fluorescence were located at various points along the chromosome; both types stained well with the Giemsa method. The different effects of various stains and pretreatments in showing up the heterochromatic segments have been compared in *Cypripedium* chromosomes [48] and in the basic rye (*Secale cereale*) karyotype [49]. In rye the positions of the heterochromatic segments are in close agreement with the distribution of the chromomeres at pachytene, including the 'gradient' of decreasing size away from the centromere. However, the segments near the centromere appear much smaller with Giemsa staining and fluorochroming than is suggested by normal chromatin dyes [4].

Several workers [50,51,52] have used Giemsa techniques to differentially stain heterochromatin associated with nucleolar organizers (NO heterochromatin). Examination of mitotic and meiotic chromosomes of *Allium cepa, Ornithogalum virens, Rhoeo discolor* and *Tradescantia edwardsiana* has demonstrated that these techniques provide a more reliable means of determining the total number and location of nucleolar organizers, even when terminal, than observations on the number of nucleoli and secondary constrictions [27].

References

[1] Feulgen, R. and Rossenbeck, H. (1924), *Hoppe-Seyl Zeit.*, **135**, 203-248.

[2] Woods, R.A. (1973), *Biochemical Genetics,* Chapman and Hall, London.

[3] Ris, H. and Kubai, D.F. (1970), *Ann. Rev. Genet.,* **4**, 263-294.

[4] Lima-de-Faria, A. (1952), *Chromosoma (Berl.),* 5, 1-68.

[5] Lima-de-Faria, A. and Sarvella, P. (1962), *Chromosoma (Berl.),* **13**, 300-314.

[6] La Cour, L.F. and Wells, B. (1971), *Cytologia,* **36**, 111-120.

[7] Ris, H. (1957), In *Chemical Basis of Heredity,* McElroy, W.D. and Glass, B. (eds.), Johns Hopkins, Baltimore.

[8] Kaufmann, B.P. and McDonald, M.R. (1957), *Cold Spring Harb. Symp. Quant. Biol.,* **21**, 233-246.

[9] Gall, J.G. (1968), *Chromosoma (Berl.),* **20**, 221-233.

[10] John, B. and Lewis, K.R. (1969), *Protoplasmatologia,* VI B, 1-125, Springer-Verlag, Wien.

[11] Jones, K. and Jopling, C. (1972), *J. Bot. Linn. Soc.,* **65**, 129-162.

[12] Kornberg, R.D. (1974), *Science,* **184**, 868-871.

[13] Noll, M. (1974), *Nature (Lond.),* **251**, 249-251.

[14] Olins, A.L. and Olins, D.E. (1974), *Science,* **183**, 330-332.

[15] Du Praw, E.J. (1966), *Nature (Lond.),* **209**, 577-581.

[16] Wolff, S. (1969), *Int. Rev. Cytol.,* **25**, 279-296.

[17] Levan, A., Fredga, K. and Sandberg, A.A. (1964), *Hereditas (Lund),* **52**, 201-220.

[18] Metzner, R. (1894), *Arch. Anat. Physiol., Physiol. Abt.* 79, 1894, 309-348.

[19] Navashin, S. (1927), *Ber. dtsch. bot. Ges.,* **45**, 415-428.

[20] Levan, A. (1946), *Hereditas (Lund),* **32**, 449-468.

[21] Östergen, G. and Andersson, L. (1973), In: *Chromosome Identification,* Caspersson, T. and Zech, L. (eds.), Academic Press, New York.

[22] Lima-de-Faria, A. (1949), *Hereditas (Lund),* **35**, 77-85.

[23] Tjio, J.H. and Levan, A. (1950), *Nature (Lond.),* **165**, 368.

[24] Bajer, A. and Mole-Bajer, J. (1969), *Chromosoma (Berl.),* **27**, 448-484.

[25] Comings, D.E. and Okada, T.A. (1971), *Exp. Cell. Res.,* **67**, 97-110.

[26] Hanzely, L. and Schjeide, O.A. (1973), *Cytobios,* 7, 147-162.

[27] Stack, S.M. (1974), *Chromosoma (Berl.),* 47, 361-378.

[28] Braselton, J.P. (1971), *Chromosoma (Berl.),* 36, 89-99.

[29] Lima-de-Faria, A. and Sarvella, P. (1958), *Hereditas (Lund),* 44, 337-346.

[30] McClintock, B. (1931), *Res. Bull. Miss. Agric. Exp. Stat.,* 163, 1-30.

[31] Dyer, R.A. (1964), *Cytologia,* 29, 155-190.

[32] Kurabayashi, M., Lewis, H. and Raven, P.H. (1962), *Am. J. Bot.,* 49, 1003-1026.

[33] Lima-de-Faria, A. and Jaworska, H. (1968), *Nature (Lond.),* 217, 138.

[34] Baumann, T.W. (1971), *Exp. Cell. Res.,* 64, 323-330.

[35] Comings, D.E. (1972), *Exp. Cell Res.,* 74, 383-390.

[36] Brown, S.W. (1966), *Science,* 151, 417-425.

[37] Pardue, M.L. and Gall, J.G. (1972), *Chromosomes Today,* 3, 47-52.

[38] Darlington, C.D. and La Cour, L.F. (1940), *Ann. Bot., N.S.,* 2, 615-625.

[39] Darlington, C.D. and Shaw, G.W. (1959), *Heredity (Lond.),* 13, 89-121.

[40] Dyer, R.A. (1963), *Chromosoma (Berl.),* 13, 545-576.

[41] Kurabayashi, M. (1952), *J. Fac. Sci. Hokkaido Univ. Ser. V.,* 6, 159-185.

[42] Woodward, J. and Swift, H. (1964), *Exp. Cell. Res.,* 34, 131-137.

[43] Pardue, M.L. and Gall, J.G. (1970), *Science,* 168, 1356-1358.

[44] Caspersson, T., Farber, S., Foley, G. *et al.,* (1968), *Exp. Cell. Res.,* 49, 219-222.

[45] Vosa, C.G. and Marchi, P. (1972), *Giorn. Bot. Ital.,* 106, 151-159.

[46] Schweizer, P. (1973), *Chromosoma (Berl.),* 40, 307-320.

[47] Vosa, C.G. (1973), *Chromosoma (Berl.),* 43, 269-278.

[48] Yamasaki, N. (1973), *Chromosoma (Berl.),* 41, 403-412.

[49] Vosa, C.G. (1974), *Heredity (Lond.),* 33, 403-408.

[50] Yunis, J.J. and Yasmineh, W.G. (1972), *Adv. Cell. Molec. Biol.,* 2, 1-46.

[51] Stack, S.M. and Clarke, C.R. (1973), *Canad. J. Genet. Cytol.,* 15, 367-369.

[52] Stack, S.M., Clarke, C.R., Cary, W.E. and Muffly, J.T. (1974), *J. Cell Sci.,* 14, 499-504.

[53] Marks, G.E. (1957), *Chromosoma (Berl.),* 8, 650-670.

[54] Haga, T. and Kurabayashi, M. (1954), *Mem. Fac. Sci., Kyushu Univ., ser. E,* 1, 159-185.

3 Chromosome division and behaviour

3.1 Chromosomal activity and the cell-cycle

Although most of the cytogenetical interest of chromosomes is focussed on their behaviour during cell division it is important to remember that the chromonemata are present throughout the whole cell-cycle, providing the physical links between cell- and organism-generations upon which the facts of genetic continuity depend (Section 1.1). The non-dividing cell has often been described as having a resting stage nucleus, because of the invisibility of the chromosomes with the usual stains, but the term interphase or, better, metabolic stage is far preferable since the chromosomes are very active, transcribing RNA for the protein synthesis involved in controlling the metabolic activity of the cell and replicating DNA prior to the next phase of chromosomal division.

In considering chromosomal activity it is necessary to distinguish between autosynthetic cells, which are engaged in active proliferation, heterosynthetic differentiating cells, and non-synthetic cells which have completed differentiation and show no or limited metabolic activity [1]. Only the former will be considered here. In actively proliferating cells the most conspicuous feature of the interphase nucleus is the synthesis of DNA prior to the next mitotic division. This occupies a relatively

short period, typically towards mid-interphase, and it is customary to recognize four stages in the cell-cycle [2] (Fig. 3.1), the mitotic (M) and DNA-synthesis (S) phases being separated by presynthetic (G_1) and post-synthetic (G_2) periods. The autosynthetic cell cycle in plants lasts for approximately 17–32 hours, of which mitosis occupies about 1.5–4 hours [1]. Of the four principal mitotic phases telophase is usually the longest, with prophase, metaphase and anaphase being progressively shorter [6].

The duration of the cell-cycle and its constituent phases is determined by feeding tritiated thymidine to actively growing roots for 30 minutes and then, by means of auto-radiography, counting the number of labelled prophases in samples taken at 2-hourly intervals [3,4]. This technique gives a double-humped curve (Fig. 3.2) from which the

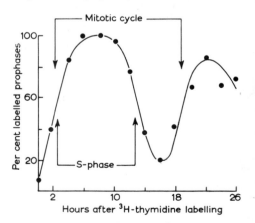

Fig. 3.2 Duration of M- and S-phases in *Nigella damascena* determined by counting labelled prophases at intervals after labelling with tritiated thymidine. (After Evans *et al.* [5]).

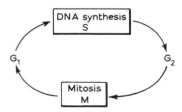

Fig. 3.1 Autosynthetic cell-cycle.

duration of the cell-cycle is shown by the interval between the 50% intercepts of the two humps in each curve, while the length of the S-phase is estimated by measuring the interval between the halfway intercepts of the ascending and descending parts of the first hump.

The duration of the complete cell-cycle is affected by general metabolic features such as respiration [7], but it has been shown [5,8,9] to be closely correlated with the nuclear DNA content, measured by means of Feulgen photometry [10]; in general, comparable tissues of different species show an increasing duration of the cell-cycle with increasing nuclear DNA. The rate of increase is around 0.3–0.38 hours per picogram of DNA and this seems to be constant for many taxa [5]. The duration of the S-phase is closely correlated with that of the cell-cycle and increases linearly with the DNA value (Fig. 3.3). Differences between taxa in the duration of the cell-cycle have been repeatedly demonstrated. It was found to be longer in dicotyledons than in monocotyledons because of the much longer G_1 period in the former [5]. This was attributed to the more

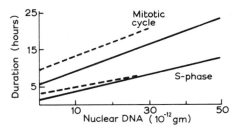

Fig. 3.3 Duration of cell-cycle and S-phase in a sample of Monocotyledones (dashes) and Dicotyledones (solid)). (After Evans *et al.* [5]).

highly coiled metaphase chromosomes of dicotyledons requiring longer to uncoil during G_1, prior to the initiation of synthesis. However, there is still a good deal of conflicting evidence on these points and several parameters are probably involved. Thus, the longer cell-cycles and S-phases in perennial than annual members of several genera of Anthemi-

deae (Compositae) [11] was not positively correlated with nuclear DNA content. It was shown that with increased nuclear DNA the heterochromatic portion of the genome increased at a faster rate than the euchromatic portion in most of the genera of Anthemideae, as had previously been demonstrated in conifers [12], and the shorter cell-cycle of annual Anthemideae may be due to a shorter S-phase resulting from the faster replication of heterochromatic DNA [13].

Autosynthetic cells, such as those considered above, can give rise to heterosynthetic and nonsynthetic cells [1] or, of major importance in cytogenetics, they can give rise to sexual, germ-line cells which eventually lead to the meiotic cycle. Many of the features of the mitotic cell-cycle are shared by meiotic cells but the latter have some specific properties. Direct observations on the chromosomes have shown that meiosis generally takes very much longer than mitosis, lasting about 6 days in *Tulbaghia* [14], and about 24 days in *Lilium* and *Trillium* [15], for example.

Since it is an unique event for any particular mega-or microsporocyte, the length of the total meiotic cell-cycle cannot be determined by the technique described for proliferating mitotic cells. The length of the cycle is greatly modified by various environmental factors, which also seem to affect other components of plant growth. Consequently, Erickson [16] utilized bud-length as an indicator of age in studies of *Lilium longiflorum* and found it more accurate than chronological data.

In general, it seems that meiosis begins earlier in relation to the inception of the cell than does mitosis and that there is no obvious G_2 period. Although the onset of the long meiotic prophase is difficult to determine precisely the S-phase seems to occur largely during later interphase, when over 99% of DNA synthesis occurs [17], or at leptotene (Fig. 3.4). Small amounts of DNA are also synthesized at zygotene and pachytene [18, 19], but these appear to be associated with chromosome pairing (Section 3.3).

In addition to the phases of DNA replication, both mitotic and meiotic cell-cycles show striking periodicity in the synthesis of RNA and the various classes of proteins. Their occurrence is profoundly important in relation to the metabolic activity of the cell and the dissemination of the genetic information conserved by the chromosomes but this is more appropriately considered in other volumes in this series. Chromosomal histone is synthesized concurrently with DNA, during which time nuclear RNA synthesis is low or absent, while synthesis of RNA and residual proteins continues beyond this point [14,20].

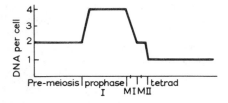

Fig. 3.4 Relative amount of DNA per cell before and during meiosis.

3.2 Chromosomal replication

During mitosis and meiosis the replication of the chromosomes, by prophase in the former and by diplotene in the latter, gives visible evidence of their suitability for providing the physical continuity throughout development of the individual and between generations required of the carriers of genetic information. With the realization of the genetical importance of DNA and the discovery that it doubles in amount prior to mitosis and meiosis (Section 3.1) it became apparent that, in some way, the two chromatids of each chromosome are the visible result of DNA replication which has occurred during interphase. Furthermore, there are obvious parallels between chromosomal replication and the semi-conservative [21] mechanism of DNA replication postulated by Watson and Crick [22], by which each new molecule would consist of one new and one old nucleotide chain.

Evidence for the semi-conservative replication of chromosomes has been found in *Vicia* [23], *Bellevalia* [24], *Crepis* [25], *Allium* [26] and *Tradescantia* [27] by means of radioisotope labelling. Seedlings of *Vicia faba* were grown in a mineral solution containing tritiated thymidine for 8 hours and then transferred to a non-radioactive solution containing colchicine, which inhibits spindle formation, for up to 34 hours. The root-tips were then squashed, stained, and autoradiographs prepared to locate the tritium incorporated into the chromosomal DNA. Cells at metaphase after 10 hours in colchicine had the normal diploid number of 12 chromosomes. Both chromatids of each chromosome were uniformly labelled, implying equal incorporation of the isotope. After 34 hours in colchicine, many cells had passed through another mitosis and at metaphase showed 24 chromosomes in each of which only one chromatid was labelled. A few cells had 48 chromosomes, indicating that they had undergone yet another mitotic division, and in these, half the chromosomes were completely unlabelled while half had one of the chromatids labelled (Fig. 3.5). Occasion-

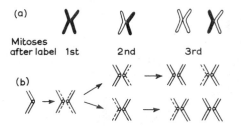

Fig. 3.5 Incorporation of tritium into *Vicia* chromosomes. (a) Labelled (black) chromatids at 1st, 2nd and 3rd metaphase after treatment. (b) Interpretation showing half-chromatids containing tritium (- - -); centromeres indicated by a circle.

ally in tetraploid ($2n = 24$) cells a chromosome would show complementary labelling in a part of each chromatid, suggesting that an exchange had occurred between sister chromatids. These results strongly support semi-conservative chromosome replication and imply that the chromatid must be composed of complementary subunits which generally remain intact, except for the points of the occasional sister-chromatid exchange. It is tempting, on such data, to consider the chromatid as a single DNA helix, but a number of observations make this unlikely.

La Cour and Pelc [28] found most X_2 chromosomes to be labelled in both chromatids, a rare occurrence reported in *Bellevalia romana* [24], in which there were also occasional chromosomes with both chromatids labelled correspondingly for a part of their length. Peacock, however, found that such iso-labelling of sister chromatids was frequent among X_2 chromosomes of *Vicia faba* and it is also known in several other organisms [27]. These results have been interpreted as indicating that there must be at least two DNA double helices per chromatid, since a synchronous replication between the component molecules would give the label segregation or unequal labelling at X_1, which would explain all the results derived from autoradiography. The direct observations of half chromatids (Section 2.1) accord with this, as do cytochemical and irradiation data indicating that the chromonemata respond as double units, each of at least a DNA duplex, before the S-phase of DNA synthesis.

The Watson and Crick model showed how DNA could replicate progressively from one end of the molecule like the opening of a zip-fastener, and there are now abundant data from prokaryotes to confirm that the genophore (*bacterial 'chromosome'*) does indeed replicate progressively in one or two directions from a single initiation point. However, there is considerable evidence from autoradiographic studies to suggest that within the eukaryote chromosome DNA synthesis is initiated at a number of different points. Thus, chromosomes exposed to tritiated thymidine for periods (pulses) which are short compared to the time required for DNA replication, incorporate the label into many sites. These data would also accord with the evidence for repeating units of DNA in chromatin (Section 2.1) and would explain how the considerable amounts of DNA in plant and animal cells can replicate much more rapidly than is expected from rates known from bacteria.

The variation in the timing of replication postulated above to account for isolabelling of sister chromatids, is also evident in the generally later replication of heterochromatin compared with adjacent euchromatin and there is often variation between chromosomes. For example, in *Scilla campanulata*, interchromosomal differences in DNA replication were directly proportional to chromosome length, while replication near the centromere was completed earlier than in more distal regions [29], but in *Phaseolus* polytene chromosomes replication begins simultaneously at several sites in the euchromatin, followed by the heterochromatin, except that near the centromere which replicates last [30]. These chromosomes also demonstrate the close control over replication, since the pattern described occurs simultaneously in each of the strands of the polytene structure. There is clearly a good deal yet to be learned about the details of chromosome replication, particularly with regard to the non-DNA components of the chromosomes, which cannot be dealt with here. The synthesis of histones during interphase, however, accords with their importance in the chromatin fibres which represent the basic structure of the chromosomes being replicated. The translation from the phase of DNA replication during interphase to the visible doubling observed at mitotic and meiotic prophase depends upon a process of spiralization and condensation. There is evidence that the proteins and possibly RNA are somehow involved in this process but the mechanism is still highly debatable, as is the extent to which other materials play a role in chromosome coiling.

3.3 Chromosome pairing and chiasmata
The precise pairing (*synapsis*) of homologous chromosomes during zygotene may begin

simultaneously at several points along the chromosome or it may be initiated at the centromere, near the telomeres or some combination of these. Synapsis is valuable for determining the degree of resemblance between the chromosomes of plants brought together in hybrid combinations but, although pairing requires homology, it has been shown in many plants that chromosomes known to be homologous fail to pair in the presence of certain genes [20], while heterochromatic regions can show non-homologous association.

Synapsis is an essential precursor for the production of the chiasmata formed between non-sister chromatids during pachytene. Although achiasmate meiosis is much more frequent in animals [20], particularly the Arthropoda, synapsis without concomitant chiasma-formation is known in, for example, *Trillium* microsporocytes [31], diploid *Lycopersicum* x *Solanum* hybrids [32] and, most convincingly, in triploid *Allium amplectans* [33].

The distribution of chiasmata does not occur at random, since the presence of one chiasma restricts the proximity of subsequent chiasmata (*positive interference*) and the presence of heterochromatin seems to restrict chiasma formation [34]. Furthermore, there is abundant evidence for genetic control of the frequency and distribution of chiasmata so that, for example, males have fewer and more localized chiasmata than female meiotic cells in *Fritillaria, Lilium* and *Allium* [35,36], while there are interchromosomal differences in *Delphinium* [37] and *Gossypium* [38] — the total recombination frequency per bivalent is apparently under strict genetic control. All critical data [20,39] support the original demonstration by Creighton and McClintock (Section 1.4) that chiasmata, involving non-sister-chromatids, are the first cytologically observed results of genetic recombination, with which they are closely correlated. Attempts to relate recombination with DNA replication (*copy choice*) have consistently failed to explain these facts, while the chiasma-type

theory, which explains the observed cytogenetical data, has lacked a credible mechanism, other than the breakage by torsion suggested by Darlington [40]. Thus, synapsis and crossing-over have for long remained one of the central enigmas of cytogenetics, but recent work on the synaptinemal complex [41] has begun to provide some of the answers.

The synaptinemal complex, first described [42] during pachytene in the crayfish *Procambarus clerkii*, has now been observed, after osmium-formaldehyde or glutaraldehyde fixation, by electron microscopic studies of meiotic prophase in a wide variety of animals and plants and it appears to be a constant feature of eukaryotic meiosis. It appears in association with leptotene chromosomes and disappears by diplotene, although elements of it may persist longer, and so is associated with the period of synapsis and chiasma-formation. The complex is best developed at pachytene, when it is universally found to consist of three parts:— two intensely staining lateral elements about 45 nm wide separated by a distance of about 100 nm through which runs a central element; most of the chromosomal chromatin lies outside the complex, flanking the lateral elements (Fig. 3.6). By means of serial sections

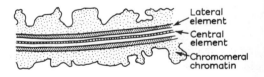

Fig. 3.6 Part of synaptinemal complex of *Phaedranassa viridiflora* at late zygotene. (From a photograph of La Cour and Wells [46]).

of pachytene nuclei it has been possible to reconstruct the complete synaptinemal complex which, in maize at least, extends along the whole length of the pachytene bivalents [43]. Furthermore, in the presence of an abnormal chromosome (K10), which is known to modify recombination between marker genes in maize, the synaptinemal complex of chromosome 3 formed a typical inversion loop (Section 4.2), while in nuclei lacking K10 the lateral components associated with chromosome 3 remained unpaired in the region of inversion heterozygosity. From these data, then, it seems that the synaptinemal complex is involved with synapsis but that its presence does not necessarily guarantee pairing.

Current views on the development of the synaptinemal complex may be summarized as follows [41]:— At leptotene in *Lilium, Tulbaghia* and other organisms, the diffuse chromatin of the chromosome (*two chromatids*) is closely associated with a single lateral element, which has been interpreted as a chromosome axis or its derivative, a structure consisting of two sister chromatid axes, or as a separate structure, identifiable by marking with metal ions, located in the split between two sister chromatids. This latter proposal would explain why the leptotene chromosome normally appears as a single structure. During zygotene the lateral elements of homologous chromosomes become roughly aligned by an unknown mechanism to lie about 300 nm apart and then, at independent points along the chromosome pair, precise pairing at a distance of about 100 nm takes place and the synaptinemal complex, as described at pachytene, is complete. The origin of the central element is not clear but it seems to be connected to the lateral elements by proteinaceous fibrils.

Distinctive DNA-protein complexes are produced at zygotene in *Lilium* [44] and it has been shown [45] that when their synthesis is blocked in *Lilium* microsporocytes, chromosome pairing and chiasma formation are prevented. Most of the chromatin of the bivalents is held outside the lateral elements of the synaptinemal complex, which suggests that only a small part of the total chromosomal content of nucleoproteins is involved in synapsis and chiasma formation [46]. That the synaptinemal complex, though necessary for synapsis, does not always result in chiasma formation has been shown in triploid *Allium amplectans* [33] which, although consistently

achiasmate, shows synapsis and the development of a complete synaptinemal complex. There is evidence [47,48] that enzymes are produced at pachytene in *Lilium* which can break and repair breaks in single strands of the DNA helix and if this is occurring in the small amount of chromatin involved in the synaptinemal complex then it is possible that problems about the precise mechanism of synapsis and chiasma formation may be somewhat closer to solution.

3.4 Chromosome segregation and meiotic drive

3.4.1 Chromosome movement
The movement of chromosomes on to the metaphase plate and their segregation to the poles at anaphase depends upon the spindle present in the cytoplasm and the kinetochore within the centromere (Section 2.2.1). The spindle, which is composed of highly orientated proteinaceous microtubules, begins to form long before the nuclear envelope breaks during prophase. Initially, the microtubules are distributed very irregularly but they become orientated, probably by breakage and growth, into the spindle which is developed along the long axis of an ellipsoidal clear zone that forms around the nucleus.

As the nuclear envelope breaks down, apparently with the assistance of the microtubules [49], the spindle fibres form a continuous system from pole to pole [50]. The kinetochores become attached to these primary fibres by kinetochore microtubules, which are developed by the kinetochore itself and/or from spindle fibres which pierce the kinetochore [49], its organizing ability being believed to depend upon the amount of repetitive DNA it contains [51]. The chromosomes are then independently pulled on to the metaphase plate by the microtubules and thence the chromosomes (meiosis-I) or daughter chromosomes (mitosis, meiosis-II) move in the same way to the poles, although in a more coordinated manner. The kinetochore plays the principal chromosomal role in this movement, but there is evidence that in long chromosomes the telomere acts as a fulcrum to give an easier and more regular separation [52]. In organisms with multiple or diffuse centromeres, the spindle fibres attach themselved at points along the length of the chromosomes, so that all parts move simultaneously polewards as rods at right angles to the spindle axis.

This, then, is how homologous chromosomes or daughter chromosomes move in a regular manner to opposite poles during cell-division to provide the physical basis of Mendelian segregation (Section 1.3). There is still a great deal to be learned about the detailed functioning of the mechanism, but it is clearly under strict, and sometimes simple, genotypic control. Thus, a single recessive allele, 'multiple spindle', in *Clarkia exilis* causes the production of 2 or more spindles in dividing pollen mother-cells [53].

As noted earlier (Section 2.2), telomeres have a very similar structure to centromeres and sometimes show non-homologous association with them (Section 3.3). In some plants they can behave as centromeres by becoming attached to and moving on the spindle during cell-division. Such 'neocentromeres' develop only in certain genotypes, but are known in a variety of plants, both at mitosis [54] and meiosis [55,56,57], although the activity is much more pronounced in the latter.

3.4.2 Meiotic drive
The chromosome behaviour described so far has concerned their exact duplication and segregation, which provides a firm and consistent basis for the transmission of genetical information. The regularity of this process is one of the unifying themes of cytogenetics. However, the normal meiotic process can be altered so that each chromosome (and gene) is not consistently present in about half of the meiotic products [58]. Such departures from Mendelian expectation, when they are not due to simple gametic or zygotic lethality, result from 'meiotic drive', which increases the frequency of alleles by the meiotic mechanism, even when opposed by selection.

Much of the work on this phenomenon has involved *Drosophila* [59], but one of the most fully explained situations involved the preferential segregation of the abnormal chromosome 10 (K10) in maize. The large terminal knob, and associated genes, on K10 are recovered normally through the pollen but in about 70% of eggs from K10/k10 heterozygotes. Other chromosomes which are heterozygous for a knob will do likewise, although only in the presence of heterozygous or homozygous K10. The knobs act as neocentromeres (Section 3.4.1), moving to the poles earlier than the centromere at both meiotic divisions, so that at second anaphase the knobbed daughter-chromosomes pass preferentially to the outer products of the linear tetrad and since one of these, the basal megaspore, is the functional egg nucleus there is excess recovery of knobbed chromosomes.

B-chromosomes (Section 5.4) show preferential segregation. Unpaired B-chromosomes in *Lilium callosum* [60], *Plantago serraria* [61] and *Tradescantia grandiflorum* [62], for example, move at random to the poles in male meiosis but they move to the micropylar pole at first anaphase in the megaspore mother-cell and are consequently preferentially included in the functional egg nucleus. In *Secale cereale* directed nondisjunction of the divided B-chromosomes occurs during first pollen mitosis, as well as in the first mitosis during the development of the 8-nucleate embryo-sac, so that preferential transmission and increase of B-chromosome frequency takes place on both male and female sides [63]. Since B-chromosomes are deleterious in large numbers (Section 5.4) the meiotic drive perhaps serves to prevent the loss of them and their positive effects in the face of selection.

3.5 Intra-plant variation
The orderly sequence of chromosome behaviour during mitosis and meiosis is designed to provide stability and genetical continuity between

cell and organism generations. However, there is abundant evidence that plants can tolerate departures from this pattern, induced by abnormal conditions, and, indeed, that this is a feature of the normal development of many species. The most common example of this is shown by cells in which, by successive replication cycles during interphase, the nuclei have two or more times as much DNA as the original nucleus. This endoreplication gives rise to polyteny in which the chromatids remain laterally attached to give ribbon-like chromosomes with up to a thousand or more strands. These polytene chromosomes are best known in the salivary glands of *Diptera* but they are characteristic of endosperm, antipodal, haustorial and other cells in the ovules of many plants.

In *Diptera* the polytene chromosomes maintain a constant morphology throughout replication but in plants, at least in the embryo suspensor cells of *Phaseolus,* they decondense and recondense during the replication cycle [65]. Endomitosis, in which the number of chromosomes increases, can also result from this replication, and endopolyploidy is particularly characteristic of non-meristematic and non-germline cells such as cortex, pith and xylem vessels, as well as in nutritive tissue like tapetum and endosperm. The reasons for the regularity of this phenomenon are not clear, although it is evident that it occurs in cells which will not give rise to new lineages exposed to selection and may indicate a breakdown of control with cell-maturity, but the association with many secretory or nutritive tissues may suggest a more positive role.

Somatic and germ-line cells characteristically have a simple relationship between their chromosome numbers (Section 3.1) but in some instances regular differences occur. The best example of this is *Rosa canina* ($2n = 35$) [66]. At meiosis 14 chromosomes form 7 bivalents and 21 remain unpaired. In megaspore mother-cells the bivalents segregate normally, while the univalents all move to the micropylar pole and divide at second anaphase so that the embryo develops from the nuclei with 28 chromosomes at the micropylar end of the linear tetrad (Section 3.4.2). In pollen mother-cells only the bivalents behave regularly; the univalents divide irregularly and segregate at random so that about 90% of pollen grains have unbalanced chromosome numbers and only those with the 7 derived from the bivalents are functional.

Mosaics or chimaeras, involving cells of different chromosome numbers, can arise irregularly in plant tissues as a result of external factors, such as wounding. In many instances, however, the cause is as yet unknown, as in a cytologically unstable plant in the F_2 progeny of the hybrid *Rubus craniensis* x *procerus* [67], which normally have $2n = 28$. When first studied, 7 months after germination, the abnormal seedling had 30 to 36 chromosomes in different root tip cells. During the next 130 days cells had from 9 to 46 chromosomes and, although the mode was always at 35 chromosomes, it was not until 168 days after the study started that all cells showed this number. Clearly, some form of canalization took place during development but neither the governing factors nor the mechanism was clear.

References

[1] John, B. and Lewis, K.R. (1969), *Protoplasmatologia VI B*, 1-125, Springer-Verlag, Wien.

[2] Howard, A. and Pelc, S.R. (1953), *Heredity (Lond.) Suppl.*, **6**, 261-274.

[3] Quastler, H. and Sherman, F.G. (1959), *Expt. Cell Res.*, **17**, 420-438.

[4] Wimber, D.E. (1960), *Amer. J. Bot.*, **47**, 828-834.

[5] Evans, G.M., Rees, H., Snell, C.L. and Sun, S. (1972), *Chromosomes Today*, **3**, 24-31.

[6] Mazia, D. (1961), *The Cell*, **3**, 77-412. Academic Press, New York.

[7] Van't Hof, J. and Wilson, G.B. (1962), *Chromosoma (Berl.)*, **13**, 39-46.

[8] Van't Hof, J. and Sparrow, A.H. (1963), *Proc. Nat. Acad. Sci. U.S.A.*, **49**, 897-902.

[9] Van't Hof, J. (1965), *Expt. Cell Res.,* **39**, 48-58.

[10] McLeish, J. and Sunderland, N. (1961), *Expt. Cell Res.,* **24**, 527-540.

[11] Nagl, W. and Ehrendorfer, F. (1974), *Plant Syst. Evol.,* **123**, 35-54.

[12] Miksche, J.P. and Hotta, Y. (1973), *Chromosoma (Berl.),* **41**, 29-36.

[13] Barlow, P.W. (1972), *Cytobios,* **6**, 55-80.

[14] Taylor, J.H. (1958), *Proc. Xth. Int. Congr. Genetics,* **1**, 63-78.

[15] Stern, H. and Hotta, Y. (1963), *Brookhaven Symp. Biol.,* **16**, 59-70.

[16] Erickson, R.O. (1948), *Amer. J. Bot.,* **35**, 729-739.

[17] Ito, M., Hotta, Y. and Stern, H. (1967), *Develop. Biol.,* **6**, 54-77.

[18] Hecht, N.B. and Stern, H. (1971), *Expt. Cell Res.,* **69**, 1-10.

[19] Sparrow, A.H., Moses, M.J. and Steele, R. (1952), *Brit. J. Rad.,* **25**, 182-189.

[20] John, B. and Lewis, K.R. (1965), *Protoplasmatologia, VI,* F1, 1-335.

[21] Delbrück, M. and Stent, G.S. (1957), In: *The Chemical Basis of Heredity,* McElroy, W.D. and Glass, B. (eds.), 699-736, Johns Hopkins Press, Baltimore.

[22] Watson, J.D. and Crick, F.H.C. (1953), *Nature (Lond.),* **171**, 737-738.

[23] Taylor, J.H., Woods, P.S. and Hughes, W.L. (1957), *Proc. Nat. Acad. Sci. U.S.A.,* **43**, 122-128.

[24] Taylor, J.H. (1958), *Genetics,* **43**, 515-529.

[25] Taylor, J.H. (1958), *Expt. Cell Res.,* **15**, 350-357.

[26] Nordqvist, T., in Lima-de-Faria, A. (1962), *Progr. Biophys.,* **12**, 282-317.

[27] Peacock, W.J. (1965), *Nat. Cancer. Inst. Monog.,* **18**, 101-123.

[28] La Cour, L.F. and Pelc, S.R. (1958), *Nature (Lond.),* **183**, 1455-1456.

[29] Evans, G.M. and Rees, H. (1966), *Exp. Cell Res.,* **44**, 150-160.

[30] Brady, T. and Clutter, M.E. (1974), *Chromosoma (Berl.),* **45**, 63-79.

[31] Parchman, L.G. and Roth, T.G. (1971), *Chromosoma (Berl.),* **33**, 129-145.

[32] Menzel, M.Y. and Price, J.M. (1966), *Amer. J. Bot.,* **53**, 1079-1086.

[33] Stack, S. (1973), *J. Cell Sci.,* **13**, 83-95.

[34] Dyer, A.F. (1963), *Chromosoma (Berl.),* **13**, 545-576.

[35] Fogwill, M. (1958), *Chromosoma (Berl.),* **9**, 493-504.

[36] Ved Brat, S. (1966), *Chromosomes Today,* **1**, 31-40.

[37] Basak, S.L. and Jain, H.K. (1963), *Chromosoma (Berl.),* **13**, 577-587.

[38] Stephens, S.G. (1961), *Genetics,* **46**, 1483-1500.

[39] Lewis, K.R. and John, B. (1963), *Chromosome Marker,* Churchill, London.

[40] Darlington, C.D. (1935), *J. Genet.,* **31**, 185-212.

[41] Westergaard, M. and von Wettstein, D. (1972), *Ann. Rev. Genet.,* **6**, 71-110.

[42] Moses, M.J. (1956), *J. Biophys. Biochem., Cytol.,* **4**, 215-218.

[43] Gillies, C.B. (1973), *Chromosoma (Berl.),* **43**, 145-176.

[44] Hecht, N.B. and Stern, H. (1971), *Exp. Cell. Res.,* **69**, 1-10.

[45] Hotta, Y., Parchman, L.G. and Stern, H. (1968), *Proc. Nat. Acad. Sci. U.S.A.,* **60**, 575-582.

[46] La Cour, L.F. and Wells, B. (1974), *Caryologia,* **27**, 83-92.

[47] Howell, S.H. and Stern, H. (1971), *J. Molec. Biol.,* **55**, 357-378.

[48] Hotta, Y. and Stern, H. (1974), *Chromosoma (Berl.),* **46**, 279-296.

[49] Bajer, A. and Mole-Bajer, J. (1969), *Chromosoma (Berl.),* **27**, 448-484.

[50] Bajer, A. (1968), *Chromosoma (Berl.),* **25**, 249-281.

[51] Steinitz-Sears, L.M. (1966), *Genetics,* **54**, 241-248.

[52] Lima-de-Faria, A. and Bose, S. (1962), *Chromosoma (Berl.),* **13**, 315-327.

[53] Vasek, F.C. (1962), *Amer. J. Bot.,* **49**, 536-539.

[54] Bajer, A. and Östergren, G. (1961), *Hereditas (Lund),* **47**, 563-598.

[55] Rees, H. (1955), *Heredity, 9*, 93-116.
[56] Hayman, D.L. (1955), *Austral. J. Biol. Sci.*, 8, 241-252.
[57] Walters, M.S. (1957), *Univ. Calif. Pub. Bot.*, 28, 335-447.
[58] Zimmering, S., Sandler, L. and Nicoletti, B. (1970), *Ann. Rev. Genet.*, 4, 409-436.
[59] Peacock, W.J. and Miklos, G.L.G. (1973), *Adv. Genet.*, 17, 361-409.
[60] Kayano, H. (1956), *Mem. Fac. Sci. Kyushu Univ.*, Ser. E, 2, 53-60.
[61] Fröst, S. (1959), *Hereditas (Lund)*, 45, 191-210.
[62] Rutishauser, A. (1960), *Heredity, 15*, 241-246.
[63] Müntzing, A. (1968), *Hereditas (Lund)*, 59, 298-302.
[64] John, B. and Lewis, K.R. (1968), *Protoplasmatologia* VI A, 1-206, Springer-Verlag, Wien.
[65] Brady, T. and Clutter, M.E. (1974), *Chromosoma (Berl.)*, 45, 63-79.
[66] Täckholm, G. (1922), *Acta Hort. Berg.*, 7, 97-381.
[67] Haskell, G. and Tun, N.N. (1961), *Genet. Res., Camb.*, 2, 10-24.

4 Chromosome changes: structure

As we have seen, the chromosomes provide the physical basis by which the genetical stability and continuity of individuals and populations is maintained. However, they are also the vehicles for generating the variation which is essential for evolution to proceed.

Regular opportunities for the generation of variability are provided by crossing-over during meiotic prophase, but less predictable changes in chromosome structure are also important for the variation and evolution of plants. These changes result from spontaneous chromosome breakage. Under normal conditions such breakage is a rare event but its frequency is subject to environmental conditions and can be increased by, for example, temperature, various high energy radiations (e.g. X-, γ- and β-rays), and some chemicals. Chromosome breakage is also higher in some hybrids [1] and it can be caused by particular genes [2]. Certain chromosomes and chromosome-segments are more likely to break than others, centromeric and other hetero-chromatic regions being particularly prone [3]. Breaks can occur at any stage of the chromosome cycle and, as noted earlier, those occurring before replication in interphase nuclei (G_1-phase) give rise to chromosome breaks, while those occurring during or after the S-phase give rise to chromatid breaks. Although these differences are important, it should be remembered that the chromatid breaks of one cell generation are the chromosome breaks at the next. Structural changes in the chromosomes depend on the behaviour of the broken ends. There is no change if they rejoin, but they may remain broken or join with other ends in the same or different chromosomes. It is customary to recognize four principal types of structural changes, deficiencies, duplications, inversions and translocations, which initially occur in the heterozygous condition since chromosomes break with a rather low frequency.

4.1 Deficiencies

Breakage without reunion leads to the piece of chromosome lacking a centromere behaving irregularly on the spindle so that it is ultimately lost from the chromosome complement and a terminal deletion results. This is not so in plants having diffuse centromeres, since the fragment can behave normally on the spindle and need not be lost, as in *Luzula* [4]. Interstitial deficiencies may result from 2 breaks in a chromosome arm and the rejoining of the ends following loss of the intervening section, but they are most commonly produced by other means, as discussed in Section 4.3.1. The cytological detection of a short deficiency is difficult or impossible but a long one can be observed directly in suitable chromosomes such as those in *Drosophila* salivary glands, or at pachytene in maize in which the absence of chromomeres or bands is evident. The close pairing of chromosomes during meiotic prophase can also be used in deficiency heterozygotes to show the location of the missing segment (Fig. 4.1).

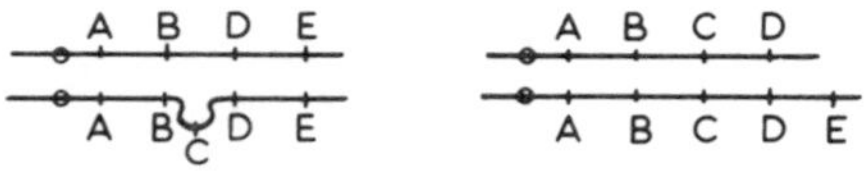

Fig. 4.1 Pachytene pairing in heterozygote for interstitial (left) and terminal (right) deficiency.

The absence of the genes lost with the deleted chromosome segment is obviously felt most in deficiency homozygotes, depending upon the number and importance of the deleted loci, but even small deficiencies are lethal in flowering plants. The haploid gametophytic stage is particularly sensitive and therefore acts as a 'sieve' for such deficiencies, though the female gametophyte, which is less independent of other tissue than the male, is more likely to tolerate them. Thus, in maize no deficiencies within the short arm of chromosome 9 were transmitted via the pollen while female gametes could function when the distal third of the short arm was lost [5].

Non-lethal deficiencies may be detectable by the decreased recombination of the genes on either side, or by the unusual phenotype to which they give rise. For example, sepal and petal colour are dramatically altered in plants of *Oenothera blandina* having deletions [6], while in maize deletions can produce effects resembling known gene mutants, such as 'white seedling', 'brown midrib' etc., due to the loss of segments containing the relevant loci [5]. This, of course, permits chromosome mapping and the correlation of cytological and genetical data. In plants heterozygous for a particular locus the loss of the chromosome segment carrying the dominant allele will allow the expression of the recessive trait. Thus, in maize the colour of the aleurone is due to the dominant allele *C*. Since aleurone nuclei are triploid a plant of constitution *Ccc* will have a coloured aleurone but cells deficient for the segment of chromosome bearing *C* are non-coloured. Since telomeres have special properties (Section 2.2.2) it has been argued that terminal deficiencies give rise to chromosomes with unstable ends, and there is much evidence for this, but broken ends are known to heal and behave as normal telomeres; the mode of breakage, as well as the tissue, appears to be important in determining whether ends will or will not stabilize [7,9].

4.2 Duplications

Under certain circumstances a chromosome complement may have additional chromosomes, which will be dealt with later (Section 5.3), or parts of chromosomes, which are termed duplications. Duplications may be on the same chromosome, on different chromosomes or, if they contain the centromere, as separate fragments.

Duplications, like deletions, can only be observed directly in favourable material such as polytene chromosomes or at pachytene in large chromosomes with distinctive chromomere patterns. The origin of a duplication was most clearly demonstrated by studies of the bar-eye phenotype in *Drosophila* [8].

Individuals with normal eyes had a chromosome segment (S) with 6 transverse bands visible on the X-chromosome in the salivary gland, while flies with narrower, 'bar' eyes having fewer facets showed the segment S to be duplicated. Amongst the progeny of 'bar' individuals could be found those with the double-bar phenotype and three S-segments in the chromosome as well as wild type flies with one S-segment. The occurrence of these types can be explained by the phenomenon of unequal crossing over (Fig. 4.2) in the female

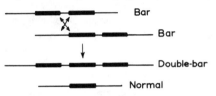

Fig. 4.2 Unequal crossing-over between duplicated regions causing bar-eye in *Drosophila*.

by which up to eight duplicated segments can be produced. Another mechanism for producing duplications has been described in maize [9] : new ends formed by breakage during meiosis can persist into the gametophyte and give rise to the so-called breakage-fusion-bridge cycle which can continue throughout the development of the endosperm. Such a cycle can result in cells of two phenotypes (Fig. 4.3),

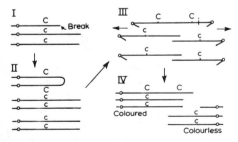

Fig. 4.3 Duplication of *C* locus of maize by breakage-fusion-breakage cycle in endosperm (3*n*). Broken end of one of 3 homologues (I) fuses with sister chromatid at prophase (II). The anaphase bridge (III) breaks to give daughter cells with genotypes *CCcc* and *cc*, respectively.

giving a mosaic of coloured and colourless tissue, depending upon the duplication or deletion of the segment bearing the colour determining allele.

Duplications tend to be less deleterious than deletions and consequently they may be expected to occur widely in natural populations. Pseudo-alleles, that is genes which behave as alleles when tested against each other but which are separable by crossing-over, are considered to result from duplication. This suggests a mechanism by which new genes can arise and, following subsequent mutation, enlarge the range of possible functions available to the organism. This has the advantage that the new variability in the duplicated segments is 'protected' by the normally functioning original segment, thus avoiding the usual deleterious effects of most mutations at single loci. Pairing between non-homologous chromosomes in haploid plants of, for example, *Antirrhinum majus* [10] and *Oenothera blandina* [11] has been considered to indicate the existence of interchromosomal duplications, but the causes of such secondary association (Section 3.3) are still obscure.

A special kind of duplication is shown by the so-called isochromosomes, which are metacentric chromosomes having the arms genetically identical (Section 5.3.2). In maize a short telocentric chromosome arose spontaneously from chromosome 5. By misdivision and reunion within the centromere this telocentric gave rise to an isochromosome which folded back at pachytene to permit the two arms to pair with each other. Duplication in isochromosomes, which usually involve a change in chromosome number (Section 5.3), is considered to have taken place naturally in a number of species, such as rye [12] and *Festuca pratensis* [13].

4.3 Inversions

When two breaks in a chromosome are followed by reunion of the new ends in reverse, the piece of the chromosome between the breakage points is rotated through 180°. The resultant reversal of the linear order of the genes was noted by Sturtevant [14], who coined the term inversion for such intrachromosomal alterations. In addition to the changed linkage arrangements of the genes, an inversion is often detectable genetically by modifications in their normal expression due to their relocation in a different part of the chromosome (*position effect*), as has been demonstrated in many organisms, including *Oenothera blandina* [15] and maize [2].

It is customary to divide inversions into two types, pericentric and paracentric, depending upon whether or not the centromere is included within the inverted segment. They differ both in their effects on chromosome morphology and in their genetical consequences.

4.3.1 Paracentric inversions

This seems to be the commonest form of inversion and is confined to a single arm of the chromosome. Although they can be detected by observing the detailed structure of suitable chromosomes, such as those in insect salivary glands, the occurrence of paracentric inversions is usually determined by studies at meiosis. Plants homozygous for an inversion show, of course, normal pairing but inversion heterozygotes form a loop at pachytene when the chromosomes pair so as to bring homologous segments together (Fig. 4.4), though in short

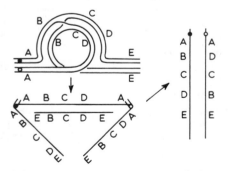

Fig. 4.4 A single cross-over within a paracentric inversion gives a dicentric bridge and an acentric fragment at anaphase. Only gametes with normal (black centromere) and inverted (open centromere) non-crossover chromosomes are viable.

inversions mechanical constraints can lead to non-homologous association.

The subsequent configuration of the chromosomes depends upon whether or not crossing-over takes place within the inverted section, the cross-over frequency depending upon the length of the inversion and its position in the chromosome in relation to the localization of the chiasmata (Section 3.3). When no crossovers take place within the inversion the chromosomes segregate normally later in meiosis, but when such a cross-over does occur a dicentric chromosome is produced which at anaphase forms the characteristic bridge accompanied by an acentric fragment (Fig. 4.4). The acentric fragment is unable to move polewards on the spindle in a normal manner and so tends to be lost from the dyad and subsequent tetrad. The chromatid bridge is usually broken by tension or severed by the developing cell walls and, wherever it breaks, the two gametes which include the products will have some genes duplicated and others deficient, and consequently they are normally inviable. Thus, only about 50% of the resultant gametes are expected to be viable and these will contain the parental inverted or normal segments. For this reason chromosome segments, now known to be inverted, were formerly considered to be cross-over suppressors when heterozygous.

In several species the expected reduction in fertility of inversion heterozygotes is not shown on the female side. In *Lilium*, *Tulipa* and maize [16,17], for example, the chromatid bridge does not break but persists into the second meiotic division. This has the effect of orientating the cross-over chromatids so that they pass to the later meiotic products in a linear sequence and, since in megasporogenesis the terminal cell of the linear tetrad develops into an ovule, it has a high chance of containing one of the parental arrangements (Section 3.4).

Where two or more cross-overs occur in paracentric inversion heterozygotes several possibilities exist (Fig. 4.5). Double crossovers within the inversion loop will give rise to completely normal gametes if they involve only 2 chromatids, 50% normal gametes if they involve 3 chromatids and no normal gametes if they involve all 4 chromatids. Further complications arise if a cross-over inside the inversion loop is associated with one between the inversion and the centromere (Fig. 4.5).

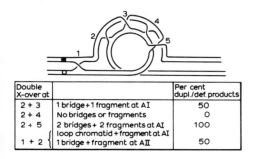

Double X-over at		Per cent dupl./def. products
2 + 3	1 bridge + 1 fragment at AI	50
2 + 4	No bridges or fragments	0
2 + 5	2 bridges + 2 fragments at AI	100
1 + 2	loop chromatid + fragment at AI	50
	1 bridge + fragment at AII	

Fig. 4.5 Effects of double cross-over within or including a paracentric inversion.

Since in most plants it is easier to observe the anaphase chromosome configurations than to interpret the pairing at pachytene, the occurrence of bridges and fragments at anaphase has been widely and often indiscriminately taken as an indication of interchange heterozygosity. However, it should be noted that in all spore mother cells in which a cross-over takes place within the inversion loop a single paracentric inversion should give a single bridge and fragment each of uniform size related to the length of the inversion. Furthermore, as can be seen from Fig. 4.5, in such cases the frequency of chromatid bridges at second anaphase is rather low and inbreeding will reduce the bridge and fragment frequency, since it lowers the number of inversion heterozygotes.

Where these criteria are not met, it has been pointed out [18,19] that bridges and acentric fragments can arise following breakage and reunion of paired chromosomes during meiosis. In rye [18] and *Paris verticillata* [20] such spontaneous breaks seem to be localized in the same region as the chiasmata but in *Tulipa*

hageri the correlation is not so evident [21]. Basically, these so-called U type exchanges can involve non-sister (NSU) or sister (SU) chromatids in a bivalent and can result from breakage of chromatids or, during mitotic prophase and pachytene-diakinesis in meiosis, half-chromatids (Chapter 2). As can be seen from Fig. 4.6, SU chromatid exchanges can result in the production of bridges and/or

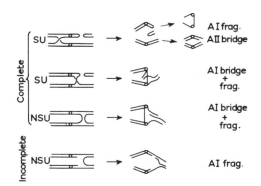

Fig. 4.6 Complete and incomplete sister and non-sister reunion after spontaneous breakage and the consequences at first and, where necessary, second anaphase. (After [19, 21].)

fragments at first and/or second anaphase, depending upon the presence and location of chiasmata, while NSU exchanges give a bridge and/or fragment at first anaphase. Of course, first anaphase bridges and fragments can persist into second anaphase, and this seems to be rather frequent, thus differing markedly from inversion bridges and fragments. Half-chromatid breakage, leading to bridges without fragments, is more difficult to determine but some of the expected configuration types have been observed in *Trillium* [22].

4.3.2 Pericentric inversions
If the two breaks involved in this type of inversion occur at equal distances from the centromere there is, as in paracentric inversions, no detectable change in the chromosome's gross morphology. However, if the breaks

occur at different distances from the centromere, the relative lengths of the arms will
change. As in paracentric inversions a loop is
formed at pachytene in plants heterozygous
for a pericentric inversion. A single cross-over
within the inversion loop also leads to an
expected 1:1 ratio of duplication-deficient

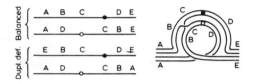

Fig. 4.7 Products of a single cross-over in a
pericentric inversion. Centromeres of normal
(black) and inverted (white) chromosome
shown.

and parental products (Fig. 4.7), but in this
case all the daughter chromosomes are centromeric and anaphase bridges and fragments are
not formed. Various sorts of double cross-
overs within the inversion loop give the balanced
and unbalanced products in the same frequencies as is expected from comparable events in
paracentric inversions but in all cases without
bridges and acentric fragments.

In conclusion, it should be noted that
although both kinds of inversions have the
disadvantage of reducing fertility when heterozygous, at least in males, their role as 'cross-
over suppressors' serves to preserve the linkage
between loci within the inverted segment and
this can be of selective advantage (Chapter 5).
Furthermore, the occurrence of inverted segments can increase the level of crossing-over
in other parts of the chromosome or else-
where in the karyotype [23].

4.4 Translocations

Reunion following breaks in non-homologous
chromosomes can lead to pieces of one chromosome being translocated on to another with
a consequent change in the linkage relationships
of the genes present on it and, depending upon
the position of the breakage points, a change in
the gross morphology of the chromosomes.
Most translocations known in plants are
reciprocal, that is there is a mutual exchange of
segments between the two non-homologous
chromosomes to give an *interchange*. 'Centric
fusions'-interchange between acrocentrics
after breaks close to the centromeres in opposite arms — are most conspicuous examples
of this but, since a change in number is usually
involved, they will be treated in Chapter 5.

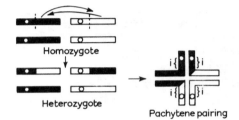

Fig. 4.8 Pachytene configuration in interchange
heterozygote following reciprocal breakage and
reunion of non-homologous chromosomes.
(i = interstitial segment).

During mitosis, or in meiosis of interchange
homozygotes, the chromosomes behave normally. However, in the heterozygous condition
meiosis shows some complications. In a plant
heterozygous for one interchange, involving a
single exchange of two non-homologous
chromosome-segments, the pairing at pachytene
brings together homologous chromosome-
segments to give a cross-shaped association of
the four chromosomes (Fig. 4.8). If pairing is
strictly homologous the centre of the cross
indicates the breakage-reunion points, but the
mechanical effects of torsion during pairing
may not permit this exactly. In the pachytene
cross it is usual to distinguish the '*interstitial
segments*' between the centromeres and the
points of exchange from the remaining *pairing
segments* of the chromosomes, since crossing-
over in the two types has different consequences.

If chiasmata are formed in each of the four
paired-arms of the cross then the four chro-

mosomes form a ring at first metaphase; progressively lower numbers of chiasmata will give a chain of four, a chain of three plus a univalent or 2 bivalents. The precise shape of the ring or chain at diakinesis and metaphase will, of course, depend upon the localization of the chiasmata and the degree to which they terminalize, as well as the length of the chromosome arms. Despite these variations, metacentric chromosomes of reasonable length with chiasmata in all arms of the interchange cross serve to show the major cytological possibilities at metaphase and their genetical consequences (Fig. 4.9). Open rings can orientate so

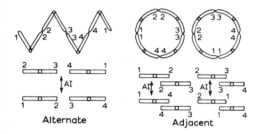

Fig. 4.9 Alternate and adjacent orientation of multiple associations at meiotic metaphase in plant heterozygous for a single interchange; anaphase products indicated.

that either adjacent homologous or adjacent non-homologous centromeres can proceed to the same pole at first anaphase: the resultant gametes are both duplicated and deficient for certain chromosome-segments and are thus usually inviable. However, a twist in the ring gives a figure-of-eight configuration so that adjacent centromeres proceed to opposite poles. Such alternate disjunction gives viable gametes with a full complement of chromosome-segments, in either the normal or translocated sequences.

Random occurrence of the three types at metaphase has been assumed to result in two thirds of the gametes being inviable. However, in maize and many other plants, only about half the gametes are sterile, which lends support to the view [24,25] that there are two

equally probable alternate disjunctions with either homologous or non-homologous centromeres moving to the same pole. Alternate disjunction is likely to be easier with increased flexibility of the ring and this will be enhanced by long, metacentric chromosomes with fewer, distal chiasmata which show a high degree of terminalization. As will be seen later (Section 6.3) these factors place constraints on the role of interchanges in the evolution of plant groups and it has been noted that the greater flexibility of chains may be the reason why they have apparently higher frequencies of alternate disjunction than comparable rings. It should be pointed out that the single interchange heterozygote just considered will show a greater frequency of unbalanced products with departure from the ring configuration, but even the latter can give abnormally high levels of sterile products if two centromeres move to opposite poles and the intervening ones, in the plane of the equator, move irregularly polewards (*discordant orientation*).

The situations described above result from crossing-over in the pairing segments only. If, in addition, crossing-over occurs in the interstitial segments the results are very different. Thus, for example, a cross-over in one of the interstitial segments in addition to crossing-over in the four pairing segments of a single inversion heterozygote gives a figure-of-eight complex at diakinesis, while if it is associated with cross-overs in the pairing segments of the other three arms, a ring of four chromosomes is produced at diakinesis (Fig. 4.10). In these instances, whether the

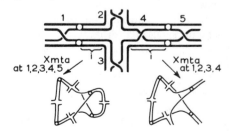

Fig. 4.10 Metaphase configurations in single interchange heterozygote with crossing-over in interstitial segment (i).

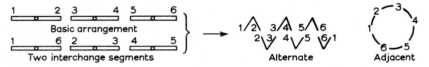

Fig. 4.11 Metaphase orientation of maximum configuration in plant heterozygous for 2 interchanges. Only alternate segregation gives balanced products.

metaphase orientation is adjacent or alternate, half the resultant gametes are inviable, since the chiasmata near the centromere cannot terminalize, so restricting the flexibility of the ring. Crossing-over in both interstitial segments gives similar genetical results, although the cytological situation is more complex [25, 26].

If a chromosome bearing a single interchanged segment, or its unchanged homologue, is involved in a further reciprocal translocation with another non-homologous chromosome a maximum configuration of six chromosomes can result (Fig. 4.11). Interchanges involving other chromosome pairs can occur to increase the size of the configuration and in some cases the whole chromosome complement can be involved, as in species of *Oenothera*, *Rhoeo discolor* and *Paeonia californica*. Crossing-over in all the pairing segments leads to rings having either adjacent or alternate orientation and, as described above, only the latter lead to balanced, viable gametes. However, where crossing-over takes place in the regions between adjacent points of interchange (*differential segments*) fertility cannot exceed fifty percent, even when the orientation is alternate.

Because of the generally low level of recombination in the differential segments they, like inverted-, duplication or deficient-segments in the heterozygous condition, tend to be inherited as units undivided by recombination. Furthermore, although larger interchange configurations have increased opportunities for orientation, resulting in unbalanced, inviable gametes, there is evidence from various plants in which they occur that a combination of equal-sized, metacentric chromosomes, distal localization of chiasmata and other features of the genotype can lead to a pre-

ponderance of alternate disjunction, as in *Campanula* [27], *Paeonia* [28], *Secale* [29] and, with almost invariable alternate disjunction, the *Oenothera biennis* group [30]. The evolutionary importance of these features will be considered in Chapter 6.

References

[1] Ehrendorfer, F. (1959), *Cold Spring Harb. Symp. Quant. Biol.*, **24**, 141-152.
[2] McClintock, B. (1950), *Proc. Nat. Acad. Sci., USA*, **36**, 344-355.
[3] Evans, H.J. and Bigger, T.R.L. (1961), *Genetics*, **46**, 277-289.
[4] Nordenskjold, H. (1961), *Hereditas (Lund)*, **47**, 203-208.
[5] McClintock, B. (1944), *Genetics*, **29**, 478-502.
[6] Catcheside, D.G. (1963), *Heredity*, **18**, 63-75.
[7] Faberge, A.C. (1959), *Genetics*, **44**, 279-285.
[8] Bridges, C.B. (1936), *Science*, **83**, 210-211.
[9] McClintock, B. (1951), *Cold Spring Harb. Symp. Quant. Biol.*, **16**, 13-47.
[10] Rieger, R. (1957), *Chromosoma (Berl.)*, **9**, 1-38.
[11] Catcheside, D.G. (1932), *Cytologia (Tokyo)*, **4**, 68-113.
[12] Muntzing, A. (1944), *Hereditas (Lund)*, **30**, 231-249.
[13] Bosemark, N.O. (1950), *Hereditas (Lund)*, **36**, 366-368.
[14] Sturtevant, A.H. (1926), *Biol. Zentralbl.*, **46**, 697-702.
[15] Catcheside, D.G. (1947), *J. Genet.*, **48**, 31-42.

[16] Darlington, C.D. and La Cour, L.F. (1941), *Ann. Bot., N.S.*, **5**, 547-562.
[17] Burnham, C.R. (1962), *Discussions in Cytogenetics*, Burges Publ. Co., Minneapolis.
[18] Rees, H. and Thompson, J.B. (1955), *Heredity* **9**, 399-407.
[19] Lewis, K.R. and John, B. (1966), *Chromosoma (Berl.)*, **18**, 287-304.
[20] Haga, T. (1953), *Cytologia (Tokyo)*, **18**, 50-66.
[21] Couzin, D.A. and Fox, D.P. (1973), *Chromosoma (Berl.)*, **41**, 421-436.
[22] Wilson, G.B., Sparrow, A.H. and Pond, V. (1959), *Amer. J. Bot.*, **46**, 309-316.
[23] Stephens, S.G. (1961), *Genetics*, **46**, 1483-1500.
[24] Burnham, C.R. (1956), *Bot. Rev.*, **22**, 419-552.
[25] John, B. and Lewis, K.R. (1965), *Protoplasmatologia*, VI, F1, 1-333.
[26] Lewis, K.R. and John, B. (1963), *Chromosome Marker*, Churchill, London.
[27] Darlington, C.D. and Gairdner, A.E. (1937), *J. Genetics*, **35**, 97-128.
[28] Walters, J.L. (1942), *Amer. J. Bot.*, **29**, 270-275.
[29] Hrishi, N.J. and Muntzing, A. (1960), *Hereditas (Lund)*, **46**, 745-752.
[30] Cleland, R.E. (1957), *Proc. Int. Gen. Symp. 1956*, *Cytologia (Suppl.)*, 5-19.

5 Chromosome changes: number

A survey of chromosome numbers in flowering plants [1,2] reveals that somatic cells can contain from 4 to 264 chromosomes. Whilst genera and even families may have the same chromosome number, in many related groups of plants the species, and sometimes populations, have different chromosome numbers. In some instances the differences involve one or two chromosomes, while in others they involve multiples of a number, usually between 7 and 13, which is found in the gametes of some of the species and which is considered to comprise the basic chromosome complement (x) or *genome* of the group. Since meiosis involves a halving of the chromosome number (Chapter 1), the products are said to be haploid (n), while between zygote formation and the next meiotic division the cells are diploid ($2n$). Plants having the basic number (x) for the group in their gametes will have two homologous complements in the zygote and derived cells, in which case $2n = 2x$. In other instances, however, there may be three ($2n = 3x$) or more basic sets in the somatic cells. Chromosome complements in which some chromosomes occur more frequently than others are known as *aneuploid*, although the term *dysploidy* is sometimes used when related organisms have a series of chromosome numbers that are not polyploid, e.g. $n = 5, 6, 7, 9$ etc. Cells with every chromosome of the basic complement represented the same number of times are known as *euploid*; the presence of more than two basic complements gives us *polyploidy*.

5.1 Polyploidy
Two extreme kinds of polyploids are generally recognized, differing in their origin and, fre-

quently, their behaviour. Autopolyploids have three or more homologous genomes, while allopolyploids are derived from two or more non-homologous genomes. However, there are many intermediate conditions so that plants can be autopolyploid for some chromosomes or chromosome-segments and allopolyploid for other chromosomes or segments, thus constituting a range of autoallopolyploids or segmental allopolyploids [3] (Fig. 5.4).

5.1.1 Autopolyploidy

Failure at mitosis or meiosis can result in autopolyploidy. Syndiploidy, in which either chromosomes divide more rapidly at mitosis than the cell wall develops between the daughter complements or in which the spindle fails during anaphase, can give rise to cells which may then divide normally to produce a tetraploid sector in an otherwise diploid plant. If this sector gives rise to flowering tissue the pollen and ovules will be diploid. Alternatively, restitution nuclei, formed at either meiotic division or during post-meiotic mitosis by the failure of the products to be enclosed by separate nuclear membranes, will also give rise to unreduced gametes. Such events occur spontaneously or can be induced by the use of spindle-inhibiting drugs such as colchicine. The union of an unreduced gamete, usually the egg, with a normal haploid gamete will give an autotriploid $(2n = 3x)$, while fertilization involving two unreduced gametes will produce an autotetraploid $(2n = 4x)$. Furthermore, diploids crossed with tetraploids will give rise to triploids. Higher polyploids can arise by similar means.

In artificially induced autopolyploids, especially tetraploids, it has been demonstrated [4, 5] that they may differ from their diploid progenitors in a number of features, such as larger cell size (obvious in pollen and stomata), slower growth rate, delayed flowering and thicker leaf-epidermis. However, it is dangerous to generalize since the features of the autopolyploid depend greatly upon those of the original diploid [3] and the genotype can modify or even nullify these features.

The genetic consequences of autopolyploidy depend to a considerable degree on the fact that there are three or more sets of homologous chromosomes and, of course, of the gene loci they bear. In meiosis, any homologue can pair with any other, but only two may synapse in any chromosome segment, although the homologues may differ in different segments. Thus, if chiasmata occur between all homologous chromosomes a multiple association can result, but their production is not invariable. An autotriploid, for example, can produce a maximum association of three, a trivalent, for each homologous set of chromosomes. At metaphase the trivalents can independently orientate in several ways (Fig. 5.1).

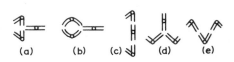

Fig. 5.1 Orientation of trivalents at metaphase. (a, b) indifferent; (c) linear; (d, e) convergent.

Numerically balanced gametes, respectively x and $2x$, can only be produced in the extremely rare event that all multivalents orientate as (d) and (e) in Fig. 5.1 to give a 2:1 segregation to the same poles. A bivalent and a univalent will result if one of the chromosomes does not form appropriate chiasmata or if the small size of the chromosomes precludes complete pairing; the latter can result in all the homologues forming univalents.

Univalents may divide either at first or second anaphase (but not normally at both), thus giving irregular segregation, or they may lag on the spindle and be excluded from the main nuclei. Consequently, whatever the meiotic configurations, triploids normally give unbalanced gametes, with numbers ranging from $x + 1$, to $2x - 1$, and so are usually sterile. Rarely, some process akin to preferential segregation (Section 3.4) increases the expected frequency of balanced gametes; in triploid *Datura*, for example, 2.6% of pollen grains have x and 1.2% have $2x$ chromosomes [6], while triploid *Anthoxanthum ovatum* has *x* chromosomes in 12.3% of pollen grains and a pollen fertility only 10–20% less than that of the diploid [7].

Autotetraploids, with four sets of homologous chromosomes can, again depending upon chromosome size and chiasma frequency, produce five types of association at meiosis — quadrivalent, trivalent + univalent, two bivalents, bivalent + two univalents and four univalents. The trivalents and univalents behave as in triploids while the quadrivalents can orientate in various ways (Fig. 5.2), giving 2:2 segregation of homologues only with parallel and convergent orientations. Although such equal segregation is more likely in auto-

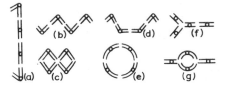

Fig. 5.2 Orientation of quadrivalents at metaphase. (a) linear; (b, c) convergent; (d, e) parallel; (f, g) indifferent.

tetraploids than in uneven-numbered polyploids, they tend to be of reduced fertility, presumably because the 2*x* gametes have unbalanced complements, so that tetraploid *Lycopersicon* [8], for example, can have 60–80% of 2*x* pollen grains but rarely more than 20% pollen viability. However, artificial autotetraploids in *Ehrarta* have 75–80% fertility [9], while in *Anthoxanthum odoratum* they produce 66.7% 2*x* pollen with 86.2% fertility, thus emphasizing the importance of the ancestral diploid genotype. A comparable role for the tetraploid gene complement is demonstrated by selection leading to increased fertility in autotetraploid *Secale* [10] and *Dactylis* [11], respectively with and without improved chromosomal segregation. Higher autopolyploids follow much the same pattern as those described above except that the number of possible associations and orientations of the larger multivalents increases very greatly.

In autopolyploids each gene locus is present as many times in somatic nuclei as the number of homologous genomes so that inheritance is more complicated than in the diploid. Thus, a locus with 2 alleles will lead to 3 possible genotypes (*AA, Aa, aa*) in the diploid but to five in a tetraploid — *AAAA* (quadruplex), *AAAa* (triplex), *AAaa* (duplex), *Aaaa* (simplex) and *aaaa* (nulliplex). Furthermore, while the diploid heterozygote produces *A* and *a* gametes in predictable frequency this is not so for the tetraploid, in which segregation can occur at either first or second meiotic division thus making it very difficult to predict the frequencies of the three gametes *AA, Aa* and *aa*. Higher levels of polyploidy and consideration of more loci quickly lead to a very complex pattern of inheritance, although the concepts involved do not differ from those of Mendelian segregation and linkage in diploids. Finally, it should be noted that selfing polyploids will not lower the level of heterozygosity as rapidly as in diploids (see also Section 5.3).

5.1.2 Allopolyploidy

Interspecific hybrids between diploid plants can show all gradations between complete chromosome pairing at meiosis and slightly reduced fertility (e.g. *Silene album* x *S dioicum*) or complete pairing and low fertility (e.g. *Festuca pratensis* x *Lolium perenne*) to low pairing and complete sterility (e.g. *Nicotiana glutinosa* x *N. sylvestris*). Where sterility is due to irregular segregation of unpaired chromosomes and not to genotypic imbalance doubling the hybrid chromosome complement will provide each chromosome with a homologue, thus allowing normal pairing and segregation and restoration of fertility. Plants with such autosyndetic pairing between chromosomes derived from the same species will behave as diploids and are, consequently, often termed amphidiploids. The classical intergeneric cross [12] between *Raphanus sativus* and *Brassica oleracea* illustrates the situation (Fig. 5.3). In the F$_1$ hybrid the 9 *Raphanus* chromosomes

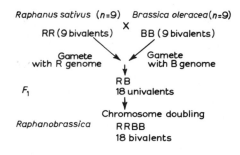

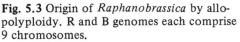

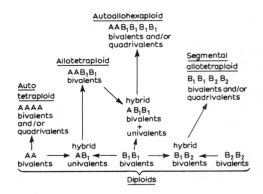

Fig. 5.3 Origin of *Raphanobrassica* by allopolyploidy. R and B genomes each comprise 9 chromosomes.

Fig. 5.4 Genomic constitution, meiotic configurations and interrelationships of various kinds of polyploids. (After Stebbins [31.])

(R genome) did not pair with the 9 *Brassica* chromosomes (B genome) and sterility was complete. Occasionally, unreduced gametes were formed and, after selfing, these gave rise to tetraploids in the F_2 generation. Such plants were fertile, formed bivalents at meiosis and had the genomic constitution RRBB. Genetic segregation was also found to be consistent with their amphidiploid status.

Where there is some pairing (allosyndesis) in the diploid hybrid the chromosomes involved are said to be homoeologous. Allopolyploids with some degree of homoeologous pairing will tend to form a proportion of multivalents, involving homologous and homoeologous chromosomes, with a consequent reduction in fertility. The hybrid between *Primula floribunda* ($n = 9$) and *P. verticillata* ($n = 9$) is completely sterile but meiotic pairing is generally good. Doubling the chromosome number gives the tetraploid *P. kewensis* ($n = 18$), which forms bivalents and some multivalents or univalents and is virtually as fertile as the parents [13].

The meiotic behaviour and fertility of segmental allopolyploids (Fig. 5.4) such as *P. kewensis* lie somewhere between those of autopolyploids and amphidiploids, depending upon the degree of homoeology. They may, however, form only bivalents as a result of the genetical control of pairing, which has been elegantly demonstrated in wheat [14]. *Triticum aestivum* is an allohexaploid ($2n = 6x = 42$)

with its genomic constitution (AABBDD) derived from the genomes of three diploid species, probably *T. monococcum* (A), *Aegilops speltoides* (B) and *A. squarrosa* (D). Only bivalents are normally formed at meiosis but if chromosome V of the B genome is missing, homoeologous pairing results, giving multiple associations, such as rings of six and chains of four chromosomes; genes in the long arm of chromosome *V* have been shown to prevent homoeologous pairing. It is possible that such genic control has been important in improving pairing, and consequently fertility and evolutionary potential, in many naturally occurring segmental allopolyploids, while differentiation of homoeologous chromosomes by, for example, structural changes will increase bivalent formation, as it could also in autopolyploids (Chapter 6).

5.2 Haploidy

In flowering plants, haploidy is the normal state in the gametophytic generation, but not in the 'diplophase'. However, natural haploid sporophytes have been observed in almost 100 species belonging to 16 families of angiosperms [15, 37].

Low frequencies of haploids are also produced experimentally by pollinating irradiated flowers with normal pollen, but one of the most successful techniques, developed during the past 10 years, involves anther culture [73, 74]. Anthers are excised just before, during or immediately after microspore mitosis and cultured on nutrient agar media containing sucrose, mineral salts and, depending on the species, various vitamins, auxins and amino acids. Such cultures can give rise to a high frequency of haploid callus tissues or embryoids by division of the vegetative cell in the microspore; the generative cell has no, or only a vestigial, role in most instances, but there is evidence that it can give rise to embryoids in *Nicotiana.* Embryoids develop directly into plantlets, which are also induced from calluses by transferring them to appropriate 'regenerative' media. The resultant plants need to be carefully screened cytologically, since many are diploid, triploid, tetraploid or hexaploid, those produced from calluses being particularly prone to chromosomal change. The production of haploid plants by this means, discovered in *Datura innoxia,* has now been tried on over 50 species, but its effectiveness varies greatly, depending upon the group, and genotypic factors are important. The technique has been intensively studied in tobacco, rice and other cereals, since it has considerable importance for plant breeding. In such haploids not only can gene action, including the study of induced mutations, be observed without the masking effect of dominance but, following diploidization induced by colchicine or resulting from endomitosis in callus tissue, homozygous, stable and true-breeding plants can be produced in a single generation.

Haploids seem particularly prevalent in hybrids, especially those between distantly related species. Perhaps the egg cell is stimulated to develop even though the gametes are too discordant for successful fertilization, or fertilization of one egg cell may stimulate the development of another in the same or an adjacent embryo sac [16]. In crosses between *Hordeum bulbosum* and both *H. vulgare* and *Triticum aestivum,* a high frequency of haploids results from the elimination of the *H. bulbosum*

chromosomes during meiosis, but the mechanism needs further study [17, 18].

5.2.1 Monohaploidy

Haploid plants of diploid species will have only one representative of each chromosome in the basic complement (x) and consequently would be expected to have irregular meiosis with no bivalent formation. This is the case, for example, in haploids of *Crepis capillaris, Datura stramonium* and *Lycopersicon esculentum.* In other instances, however, some chromosome pairing does occur, as in *Antirrhinum majus, Hordeum spontaneum, Oenothera blandina, Oryza glaberrima* and *Secale vulgare,* but this seems to be due to non-homologous secondary association and claims for chiasma formation are doubtful [16]. The univalents either divide at first anaphase or pass at random to the poles; the chance of obtaining haploid gametes is low and monohaploids are usually sterile.

5.2.2 Polyhaploidy

Sporophytic haploids of this type, derived from polyploids, may contain homoeologous genomes giving more meiotic pairing than in monohaploids. Thus, polyhaploid *Triticum aestivum* has three times as much pairing as the related monohaploid *Aegilops longissima* [19]. Complete synapsis was observed in polyhaploids in species of *Capsicum, Dactylis, Medicago, Solanum, Valeriana, Bromus, Parthenium* and *Sorghum,* the last three of which were self-fertile [15]. Clearly, therefore, polyhaploids can produce fertile gametes following chromosome pairing and segregation in the same way as autopolyploids, and polyhaploids of true amphidiploids should behave exactly like monohaploids. Different levels of homoeology will produce different results and there can be considerable variation even within polyhaploids of the same parentage; polyhaploid *Solanum polytrichon* ($2n = 24$), for example, can form 4–12 bivalents, though the mean is about 8, but it is always sterile [20].

5.3 Aneuploidy

Aneuploid changes in chromosome number, involving less than the whole genome, may result from addition or subtraction of complete chromosomes or from the gain or loss of centromeres without significant change in the total amount of genetic material. Plants, such as *Luzula*, with polycentric chromosomes can increase the chromosome number by fragmentation (*agmatoploidy*) without changing either the total chromosomal material or the number of centromeres in the cell.

5.3.1 Aneuploidy involving whole chromosomes

In diploids nondisjunction of a chromosome during mitosis will lead to both sister chromatids passing to the same pole so that one daughter cell will have three representatives and the other cell one representative of that chromosome. Nondisjunction at meiosis will lead to similar unbalanced gametes and progeny, as will unequal segregation in autopolyploids, especially autotriploids, and interchange heterozygotes (Section 4.3). Cells, tissues or plants lacking one chromosome are *monosomic* and those lacking both homologues are *nullisomic*. Trisomics have three, and tetrasomics four, representatives of a chromosome, while two different extra chromosomes give a double trisomic, and so on. The loss of chromosomes is usually much more deleterious than their addition so that monosomics and nullisomics are rare in diploid organisms but, because of the buffering effects of the additional genomes, they are much more likely to occur in polyploids. Indeed, all 21 possible nullisomics and monosomics have been recorded in *Triticum aestivum* [21], but even here the former have poor vigour and low fertility. All possible monosomics have been established in *Nicotiana tabacum* [22], as have many in *Avena sativa* [23] and *Gossypium hirsutum* [24]. Occasional monosomics have been reported in the diploids *Datura stramonium* [25], *Lycopersicon esculentum* [26], *Nicotiana alata* [27] and *Zea mays* [28]. Only two primary monosomics of tomato [26] were viable, these had frequent non-homologous

pairing at pachytene and the monosomy was not transmitted to the very large progenies tested. Trisomics and tetrasomics are common and most data are on the former. Trisomics are usually less vigorous and fertile than the diploid and they frequently have distinctive phenotypes. Thus, in trisomic *Datura stramonium* ($n = 12$) twelve different atypical capsule types are known, depending upon which chromosome is in triplicate [29], and comparable morphological effects occur in trisomic *Lycopersicon* [30]. At meiosis the chromosomes causing the aneuploidy will behave as in the autopolyploids or haploids described earlier. Thus, trisomics form either trivalents or a bivalent and a univalent, tetrasomics can form quadrivalents, and so on. The trisomics discussed so far are primary trisomics, with the three chromosomes completely homologous. However, in *Datura* Blakeslee also identified secondary trisomics, with the extra chromosome being an isochromosome (Section 4.1) derived from one of the original pair, and tertiary trisomics, in which the extra chromosome involves parts of two non-homologous chromosomes present in the diploid complement (Fig. 5.5).

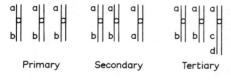
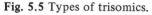

Fig. 5.5 Types of trisomics.

In trisomics the extra chromosome tends to be lost, the frequency of transmission being partly a function of its length and hence of the chance of its being associated with its two homologues by chiasmata [28]. In *Lycopersicon* 0.44–24.7% of functional eggs have $x + 1$ chromosomes and in *Datura* [31] 10–32%, depending upon the chromosome involved [30]; pollen shows similar transmission differences. Plants of *Spinacia oleracea* ($n = 6$) trisomic for chromosome 4, having a 'curled' phenotype with stunted and twisted leaves, were found to revert to normal. This was due to the elimination of one of the three representatives of chromosome 4 and occurred at various stages during plant development to give normal and curled sectors of different sizes. Such chimaeras, involving a change from $2n = 13$ to $2n = 12$, could involve a single leaf or different shoots on the plant [32]. The genes borne on the chromosomes causing the aneuploidy will segregate differently from those carried by the other chromosomes and this can be used to discover which chromosome carries a particular locus. As in autopolyploids, heterozygosity of genes on the trisomic and tetrasomic chromosomes is maintained longer under inbreeding than when only two homologues are present [33,34].

5.3.2 Aneuploidy involving centromeres only

Gain or loss of centromeres without significant change in the total amount of genetic material is particularly important in evolution (Chapter 6), since they can change the linkage relationships of the genes borne on the chromosome segments involved and, at the diploid level, result in changes in the basic number (x) upon which polyploidy can build.

Gain of centromere. Misdivision of the centromere (*centric fission*), first postulated by Darlington [35], was given support by the demonstration of its basically duplicate nature [36]. Although the exact mechanics of the division of the centromere need further clarification in view of recent observations on its ultrastructure (Section 2.2.1), there is considerable evidence that a metacentric chromosome can, by transverse division of the centromere, give two telocentrics (Fig. 5.6). Telocentrics may be abnormal, commonly showing nondisjunction and misdivision when they are univalent, but whether this is a consequence of an incomplete centromere [35] or because they tend to arise in material cytologically abnormal in other respects [38], is not clear. Certainly, such instability was demonstrated in some plants of *Campanula persicifolia* [39], in which one of the two telocentrics gave rise

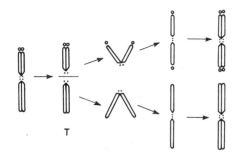

Fig. 5.6 Diagram to show origin of telocentric chromosomes (T) and/or isochromosomes by centromeric breakage. (After Strid [40].)

to an isochromosome; both telocentrics and derived isochromosomes were irregularly distributed at mitosis and meiosis, and liable to further misdivision and fragmentation. However, stable telocentric chromosomes do occur in *Campanula persicifolia* [39] and are known, for example, in *Nigella doerfleri* [40], *Oxalis dispar* [41], *Tradescantia micrantha* [42] and *Zebrina* sp [43], while stable isochromosomes are found in *Nicandra physalodes* [44].

Nigella doerfleri normally has $2n = 12$, but a population on the Greek island of Ios contained plants with $2n = 12$ and 14, those with $2n = 14$ having a pair of metacentrics replaced by two pairs of telocentrics (Fig. 5.7). Artificial

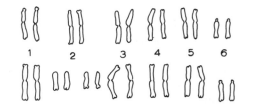

Fig. 5.7 Karyotypes of *Nigella doerfleri*. (Above): normal complement (10 metacentrics, 2 acrocentrics). (Below): derived complement (8 metacentrics, 4 telocentrics, 2 acrocentrics). (After Strid [40].)

hybrids ($2n = 13$) regularly formed five bivalents and a trivalent at meiosis, the latter resulting from the pairing of the two telocentric chromosomes with the arms of the metacentric

43

chromosome, and anaphase separation was normal. On selfing, the hybrid produced 12-, 13- and 14-chromosome plants in the ratio of 1:2:1, indicating random fusion of 6- and 7-chromosome gametes, and demonstrating that division of a metacentric to give two telocentric chromosomes, which would originally arise in the heterozygous condition, can lead to structural homozygosity, at least in a self-fertilizing species such as this [40].

Apart from their instability, the establishment of telocentric chromosomes depends upon possible genetic consequences of their formation. The greater number of linkage groups can lead to greater variability and centric fission might also separate the tightly linked genes in the pericentric region where recombination is usually low in metacentric chromosomes [42]. Centric fission could be of selective advantage if one of the resultant telocentric chromosomes contained a number of deleterious genes and was lost [41] and it has been suggested [45] that this process might be involved in the formation of supernumerary chromosomes, many of which seem to be telocentrics or isochromosomes (Section 5.4). As will be noted later (Section 5.4) many plants have irregularly occurring small B chromosomes in addition to the standard, stable complement. These B chromosomes are often largely heterochromatic and are frequently reduced to an almost naked centromere. It has often been proposed [46,47] that translocation of segments of the standard chromosomes to these largely inert B chromosomes would, at one step, convert them from non-essential to essential members of the complement. Although it has proved experimentally possible, for example, to translocate part of chromosome 4 to a B chromosome in *Zea mays* [48] there seems to be little or no clear evidence of such a mechanism operating in nature.

Loss of centromere. Since misdivision of the centromere can give two telocentric chromosomes in place of one metacentric it might seem likely that the reverse process could

occur, with the centromeres of two telocentrics fusing to give a reduction in chromosome number. As stable centromeres, like telomeres, do not seem able to fuse (Section 2.2), such a centric fusion would require a break along the mid-point of each centromere permitting true whole-arm interchange [49]. However, there is no clear example that this has occurred and most cases of supposed centric fusion appear to involve a so-called Robertsonian translocation between two acrocentrics (Fig. 5. 8). Thus, breaks in the short arm of one acrocentric and in the long arm of the other, both very close to the centromere, followed by interchange will result in a long metacentric

Fig. 5.8 Derivation of long metacentric and very short metacentric (often lost) chromosomes from non-homologous acrocentrics by Robertsonian translocation.

and a very short metacentric chromosome, which can be little more than a centromere. Although, the small chromosome may persist, as in *Haplopappus* [50] and *Haworthia* [51], it is usually lost because it cannot form chiasmata in a regular way. If the small chromosome carries little genetic material its loss will not be too deleterious, and this seems to be borne out by the many examples in natural populations of reduction in chromosome number attributable to Robertsonian translocation. The Robertsonian relationship between the original and the derived chromosome complements can be shown by the meiotic configuration in the heterozygote. Thus, in the classical hybrid between *Crepis neglecta* (2n = 8) and its derivative *C. fuliginosa* (2n = 6) [52], two non-homologous chromosomes of the former species paired with only one chromosome of the latter, indicating that most of the material in the B and C chromosomes of *C. neglecta* has been united by interchange into the B chromosome of *C. fuliginosa*. The pairing of other chromosomes in the hybrid (Fig. 5.9) suggests that the A and D

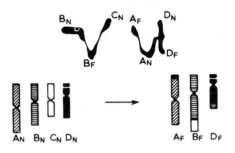

Fig. 5.9 (Above) Meiotic pairing in hybrid *Crepis fuliginosa* x *neglecta*. (After Tobgy [52]). (Below) Deduced arrangement of chromosome segments in *C. neglecta* (left) and derived *C. fuliginosa*.

chromosomes of each species have diverged by unequal translocation and, perhaps, inversion.

There is evidence for a similar reduction in chromosome number within *Haplopappus gracilis* [53]. A population in southern Arizona comprised plants with $2n = 4,5$ and 6, those with $2n = 5$ being hybrids between the other two types and showing a bivalent and a trivalent at meiosis. The morphology of the somatic chromosomes and meiotic pairing in natural and synthesized hybrids (Fig. 5.10)

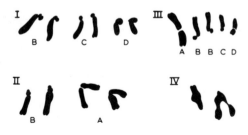

Fig. 5.10 *Haplopappus gracilis*. Somatic karyotypes of 6-chromosome (I) and derived 4-chromosome (II) race. Karyotype (III) and meiotic metaphase configuration (IV) of artificial hybrid.

suggested that *H. gracilis* with $2n = 4$ was derived by Robertsonian translocation from plants with $2n = 6$, which, incidentally, had themselves been derived by centromeric reduction from *H. ravenii* ($2n = 8$) or a close relative [50].

A more complex situation has been demonstrated [49] in *Gibasis*, which contains diploid ($2n = 10$) and autotetraploid ($2n = 16$) cytotypes. Single chromosome sets (x) of these plants respectively consist of 2 metacentrics + 3 acrocentrics and 3 metacentrics + 1 acrocentric (Fig. 5.11). In artificial F_1 hybrids acrocentrics pair with metacentric arms thus confirming the Robertsonian relationship between the two basic complements. Since the derived complement ($x = 4$) is present in the autotetraploid there are two homologous metacentrics able to pair with the two acrocentrics of the diploid ($x = 5$) in various combinations, most frequently giving trivalents but also sometimes quadrivalents (Fig. 5.11).

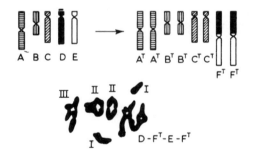

Fig. 5.11 Haploid karyotypes of diploid and derived autotetraploid after Robertsonian translocation in *Gibasis schiedana*, and meiotic metaphase in artificial hybrid showing quadrivalent including acrocentrics and derived metacentrics. (After Jones [49].)

5.4 Supernumerary chromosomes
The standard chromosome complement, composed of two or more basic genomes, is seen to be a rather delicately balanced system which is easily upset by the addition or, particularly, subtraction of chromosomes or chromosome-

segments. Among other things mitosis and meiosis are designed to maintain this balance and all the chromosomes are involved in this. However, other chromosomes of irregular occurrence, which did not behave normally in cell division and which were morphologically distinct, were first noted in maize [54], and given the name B-chromosomes [55] to distinguish them from the A chromosomes of the standard complement. Supernumerary, accessory or B-chromosomes, to mention just three of the terms used, are known in a wide variety of plants belonging to at least 88 genera in 20 families, as well as in many animals including man [56].

Supernumerary chromosomes can be morphologically indistinguishable from the A-chromosomes, as in *Clarkia unguiculata* [57], *C. williamsonii* [58] and *Narcissus* [59], but they are usually smaller and more heterochromatic. In *Clarkia elegans* [60] supernumerary chromosomes have been shown to be derived from the A-complement following reciprocal translocation and they may arise *de novo* rather frequently in populations with high levels of translocation heterozygosity [58]. However, in most cases the origin of B-chromosomes is not so clear, although they have been considered to be telocentrics or isochromosomes derived from them [45]. B-chromosomes do not pair with the A-complement, although achiasmate associations between them have been observed in, for example, *Dactylis glomerata* [61].

Homologous B-chromosomes can pair and even form multivalent associations [45], the chiasma frequency being highly subject to environmental factors. B-bivalents usually orientate and segregate normally at meiosis (but see Section 3.4). Unpaired B-chromosomes, however, often show variable behaviour at first anaphase, like univalents, but they may be more predictable. Thus, in many grasses they consistently divide only at first anaphase and at second anaphase the daughter B-chromosomes lag on the spindle and are frequently excluded from the tetrad

nuclei. In other cases, such as *Anthoxanthum aristatum,* they divide only at second anaphase and segregate regularly to the tetrads.

B-chromosomes occur at highly variable frequencies in different populations, plants and even tissues within a plant. At mitosis they may divide normally or show various kinds of instability, such as nondisjunction or lagging at anaphase, resulting in their loss or unequal segregation. Supernumeraries within the same species can differ in these respects. Thus in *Brachychome lineariloba* the large (*c.* 4μm) B-chromosomes are mitotically stable, while the smaller (< 1 μm) ones, which seem to be little more than naked centromeres, are very unstable [62]. Supernumeraries are often eliminated from certain tissues at particular stages of development. In *Haplopappus gracilis* [63], for example, they are of variable occurrence in the shoot and are eliminated from the roots, in *H. spinulosum* [64] they are constant in the roots but variable in the germ-line, while in *Sorghum purpureosericeum* [65] they are retained only in reproductive organs.

Supernumerary chromosomes are frequently considered unimportant for the efficiency of the plant and their often high degree of heterochromatinization supports this view. However, studies of their occurrence and transmission in natural and experimental populations (Chapter 6), as well as their directed segregation within some tissues (Section 3.4), suggest that they are not completely inert and various data support this. Increased sowing density, and therefore increased selection pressures, reduced the survival of *Secale cereale* plants with B chromosomes but increased the survival of *Lolium perenne* plants with two B-chromosomes in comparison with individuals lacking supernumeraries [66]. In *Secale cereale* [67] and several other species B-chromosomes can increase recombination in the A-complement but in *Najas marina* [68], at least, they seem to reduce chiasma frequency. In some *Triticum speltoides* populations plants with B chromosomes have more interstitial chiasmata and fewer terminal chiasmata than plants lacking them [69], while they can greatly reduce the normally high level of homoeologous pairing in *Lolium* [70] and *Triticum* [71] hybrids. Although in some of these examples the effect of the B-chromosomes is proportional to their number, higher numbers of them generally have a deleterious effect on phenotype and fertility, while even a few supernumeraries can cause delayed maturation and reduced fertility in *Anthoxanthum aristatum* [72]. Certainly these chromosomes are not the 'nuclear parasites' they were once considered.

5.5 Sex chromosomes

Dioecism, with male and female flowers borne on separate plants, is relatively rare in higher plants. Generally, the sex organs are produced in response to developmental factors (carpels are normally more terminal than anthers), indeed, sex-expression in plants is widely susceptible to enviromental modification [75]. In dioecious species the simplest form of sex-determination, genetically, is by means of alleles at a single gene locus. Crosses between dioecious and monoecious forms of *Ecballium elaterium* suggest that there are 3 alleles at the x locus — x^M(male), x^+(monoecious) and x^f(female) — with relative dominance in that order [76]. It has been suggested [76] that, by suppression of crossing-over near the sex-determining locus after structural and genic change, differential segments can evolve as a more advanced state. Although not visible cytologically, this situation can be determined genetically by sex-linked inheritance (Section 1.3) and shows that, normally, males are heterogametic and females homogametic; in angiosperms only *Fragaria* [77] and *Potentilla fruticosa* [78] are known to have heterogametic females. Increasing differentiation of the sex chromosomes, so that pairing between them is increasingly reduced, is thought to lead to their becoming visibly different from each other and from the other chromosomes (autosomes).

Cytologically identifiable sex chromosomes are known in some angiosperms (*Cannabis, Humulus, Spinacia, Silene*) and in a number of bryophytes. At meiosis in the sporophyte of the liverwort *Sphaerocarpus donellii*, for example, the large X and the small Y-chromosomes each segregate to half the spores, those with the X-chromosome giving rise to female gametophytes and those with the Y-chromosome to males. In *Silene album* [77], which is diploid, female plants have 2 X-chromosomes and males an XY pair, with the Y clearly larger than the X. By colchicine treatment and heat shock a series of polyploids was induced and plants were produced with 2, 3 or 4 sets of autosomes, up to 5 X and up to 2 Y-chromosomes. It was found that in general the presence of the Y-chromosome gives a male and its absence a female, irrespective of the number of X-chromosomes and sets of autosomes. The only exceptions are tetraploid plants with 4 X and 1 Y-chromosomes, which are usually hermaphrodite. Chromosome fragmentation by X-rays gave plants with partly deleted Y-chromosomes and study of these plants suggests that one part of the Y-chromosome contains genes controlling anther-development and other parts have genes suppressing the development of female characters. *Silene* is thus like man and other mammals in that sex is determined by the balance between X and Y-chromosomes, a ratio of 4X:1Y being required even to produce an hermaphrodite, with the autosomes of little or no importance.

A different situation exists in *Rumex acetosa* and related species. Among the progeny of induced triploids and tetraploids were plants with several numbers of chromosomes which showed that, as in *Drosophila*, sex is determined by the balance between X-chromosomes and autosomes, the Y-chromosome being of little importance. Plants with ratios of 1X:1A (where A = 6 autosomes) were female, whether of not Y-chromosomes were also present, while plants with ratios of 1X:2 or 3A were male; intermediate ratios (3A + 2X, 4A + 3X, 6A + 4X, with or without Y-chromosomes) were intersexes. These species also demonstrate how the sex chromosome mechanism can be modified by the inclusion of one or more pairs of autosomes as a result of translocation. In *Rumex hastatulus* [79] a race from Texas has females with $2n = 10 = 2A + XX$ and males with $2n = 10 = 2A + XY$, while a race from North Carolina has females with $2n = 8 = 2A + XX$ and males with $2n = 9 = 2A + XY_1 Y_2$. In the latter, segregation of the $XY_1 Y_2$ trivalent at meiosis results in X or $Y_1 Y_2$ pollen. Hybrids show that one of the autosome pairs in the Texas race is homologous with an end of the X-chromosome and one of the Y-chromosomes in the North Carolina race, indicating that an interchange has taken place. Such translocations between autosomes and X-chromosomes have the effect of bringing into the sex chromosome mechanism autosomes (*neo-Y-chromosomes*) which can then begin to differentiate as indicated earlier. Repeated translocations can include more autosomes and in *Humulus* up to 5 chromosomes can be involved in sex determination.

In addition to size differences, X and Y-chromosomes frequently differ in their replication and chromatin. This has been much studied in mammals, where the Y-chromosome is often heterochromatic, and one of the X-chromosomes in the female becomes so during early embryogenesis. In *Rumex thyrsiflorus* [80] the X-chromosome in males and the 2 X-chromosomes in females are euchromatic and undergo DNA replication simultaneously with euchromatic autosomes. The 2 Y-chromosomes in males are heterochromatic and late-replicating, though they differ in the onset of replication.

References

[1] Federov, A. (1969), *Chromosome Numbers of Flowering Plants,* Komarov Bot. Inst., Leningrad.

[2] Moore, R.J. and Cave, M.S. (eds) (1965-1974), *Index to Chromosome Numbers,* Reg. Veg., Utrecht.

[3] Stebbins, G.L. (1951), *Variation and Evolution in Plants,* Columbia Univ. Press, New York.

[4] Randolph, L.F. (1941), *Amer. Nat.,* **75**, 347-363.

[5] Stebbins, G.L. (1941), *Amer. J. Bot.,* **28**, Suppl., 65.

[6] Satina, S. and Blakeslee, A.F. (1937), *Amer. J. Bot.,* **24**, 518-527.

[7] Carroll, C.P. (1966), *Chromosoma (Berl.),* **18**, 19-43.

[8] Dawson, G.W.P. (1962), *An Introduction to the Cytogenetics of Polyploids,* Blackwell, Oxford, Fig. 16.

[9] Stebbins, G.L. (1949), *Proc. 8th Int. Congr. Genet.,* 461-485.

[10] Roseweier, J. and Rees, H. (1962), *Nature (Lond.),* **195**, 203-204.

[11] McCollum, G.D. (1958), *Chromosoma (Berl.),* **9**, 571-605.

[12] Karpechenko, G.D. (1928), *Zeit. Ind. Abst. Vererbungsl.,* **39**, 1-7.

[13] Newton, W.C.F. and Pellew, C. (1929), *J. Genet.,* **20**, 405-467.

[14] Riley, R. (1960), *Heredity,* **15**, 407-429.

[15] Kimber, G. and Riley, R. (1963), *Bot. Rev.,* **29**, 480-531.

[16] John, B. and Lewis, K.R. (1965), *Protoplasmatalogia,* VI, F1, 102-103.

[17] Kasha, K.J. and Kao, K.N. (1970), *Nature (Lond.),* **225**, 874-876.

[18] Barclay, I.R. (1975), *Nature (Lond.),* **256**, 410-411.

[19] Riley, R. and Chapman, V. (1957), *Heredity,* **11**, 195-207.

[20] Marks, G.E. (1955), *Nature (Lond.),* **175**, 469.

[21] Sears, E.R. (1953), *Amer. Nat.,* **87**, 245-252.

[22] Clausen, R.E. and Cameron, D.R. (1944), *Genetics,* **29**, 447-477.

[23] Andrews, G.Y. and McGinnis, R.C. (1964), *Canad. J. Genet. Cytol.,* **6**, 349-356.

[24] White, T.G. and Endrizzi, J.E. (1965), *Genetics,* **51**, 605-612.

[25] Blakeslee, A.F. and Avery, A.G. (1938), *Carnegie Inst. Wash. Publ.,* **501**, 315-351.

[26] Khush, G.S. and Rick, C.M. (1966), *Chromosoma (Berl.),* **18**, 407-420.

[27] Avery, P. (1929), *Univ. Calif. Publ. Bot.*, **11**, 265-284.

[28] Einset, J. (1943), *Genetics,* **28**, 349-364.

[29] Blakeslee, A.F. (1930), *Smithsonian Rep.* 1930, 431-450.

[30] Rick, C.M. and Barton, D.W. (1954), *Genetics,* **39**, 640-666.

[31] Satina, S. and Blakeslee, A.F. (1937), *Amer. J. Bot.,* **24**, 621-627.

[32] Ellis, J. R. and Janick, J. (1959), *J. Hered.,* **50**, 272-274.

[33] Sánchez-Monge, E. (1972), *Anal. Inst. Nac. Investig. Agr., Ser. Prod. Veget.,* **2**, 11-21.

[34] Sánchez-Monge, E. (1972), *Rev. Real Acad. Cienc. Exact. Fis. Nat. Madrid,* **66**, 339-347.

[35] Darlington, C.D. (1939), *J. Genet.,* **37**, 341-364.

[36] Lima-de-Faria, A. (1956), *Hereditas (Lund),* **42**, 85-160.

[37] Kasha, K.J. (ed.), (1974), *Haploids* in *Higher Plants,* Univ. of Guelph, Ontario, Canada.

[38] Marks, G.E. (1957), *Amer. Nat.,* **91**, 223-232.

[39] Darlington, C.D. and La Cour, L.F. (1950), *Heredity,* **4**, 217-248.

[40] Strid, A. (1968), *Bot. Notiser,* **121**, 153-164.

[41] Marks, G.E. (1957), *Chromosoma (Berl.),* **8**, 650-670.

[42] Jones, K. (1969), *Chromosomes Today,* **2**, 218-222.

[43] Mattsson, O. (1964), *Abstr. 10th Internat. Bot. Congress,* 99-100.

[44] Darlington, C.D. and Janaki-Ammal, E.K. (1945), *Ann. Bot., N.S.,* **9**, 267-281.

[45] Battaglia, E. (1964), *Caryologia,* **17**, 245-299.

[46] Darlington, C.D. (1937), *Recent Advances in Cytology,* edit. 2, Churchill, London.

[47] Swanson, C.P. (1958), *Cytology and Cytogenetics,* MacMillan, London.

[48] Roman, H. (1947), *Genetics,* **39**, 365-377.

[49] Jones, K. (1974), *Chromosoma (Berl.),* **45**, 353-368.

[50] Jackson, R.C. (1962), *Amer. J. Bot.,* **49**, 119-132.

[51] Brandham, P.E. (1974), *Chromosoma (Berl.),* **47**, 85-108.

[52] Tobgy, H.A. (1943), *J. Genet.,* **45**, 67-111.

[53] Jackson, R.C. (1965), *Amer. J. Bot.,* **52**, 946-953.

[54] Longley, A.E. (1927), *J. Agric. Res.,* **35**, 769-784.

[55] Randolph, L.F. (1928), *Anat. Rec.,* **41**, 102.

[56] Muntzing, A. (1974), *Ann. Rev. Genet.,* **8**, 243-266.

[57] Mooring, J.S. (1960), *Amer. J. Bot.,* **47**, 847-854.

[58] Wedberg, H. L., Lewis, H. and Venkatesh, C. S. (1968), *Evolution,* **22**, 93-107.

[59] Fernandes, A. (1949), *Bolm. Soc. Brot. Sér. 2,* **23**, 5-69.

[60] Lewis, H. (1951), *Evolution,* **5**, 142-157.

[61] Shah, S.S. (1963), *Chromosoma (Berl.),* **14**, 162-185.

[62] Carter, C.R. and Smith-White, S. (1972), *Chromosoma (Berl.),* **39**, 361-379.

[63] Östergren, G. and Frost, S. (1962). *Hereditas (Lund),* **48**, 363-365.

[64] Li, N. and Jackson, R.C. (1961), *Amer. J. Bot.,* **48**, 419-426.

[65] Darlington, C.D. and Thomas, P.T. (1941), *Proc. Roy. Soc. Lond. B.,* **859**, 127-150.

[66] Hutchinson, J. (1975), *Heredity,* **34**, 39-52.

[67] Jones, R.N. and Rees, H. (1967), *Heredity,* **22**, 333-347.

[68] Viinikka, Y. (1973), *Hereditas (Lund),* **75**, 207-212.

[69] Zarchi, Y., Hillel, J. and Simchen, G. (1974), *Heredity,* **33**, 173-180.

[70] Evans, G.M. and Macefield, A.C. (1973), *Chromosoma (Berl.),* **41**, 63-73.

[71] Dover, G. A. and Riley, R. (1972), *Nature (Lond.),* **240**, 159-161.

[72] Östergren, G. (1947), *Hereditas (Lund),* **33**, 261-296.

[73] Sunderland, N. (1974), In *Haploids in Higher Plants*, Kasha, K.J. (ed.), Univ. Guelph, Ontario, pp. 91-122.

[74] Vasil, I.K. and Nitsch, C. (1975), *Zeits. f. Pflanzenphysiol.*, **76**, 191-212.

[75] Heslop-Harrison, J. (1957), *Biol. Rev.*, **32**, 38-90.

[76] Galan, F. (1950), *Proc. VII Int. Congr. Bot.*, *Stockholm*, p. 340.

[77] Westergaard, M. (1958), *Adv. Genet.*, **9**, 217-281.

[78] Grewal, M.S. and Ellis, J.R. (1972), *Heredity*, **29**, 359-362.

[79] Smith, B.W. (1972), *Chromosomes Today*, 2, 172-182.

[80] Žuk, J. (1972), *Chromosomes Today*, **2**, 183-188.

6 Chromosomes and plant evolution

The student of evolution is basically looking at the diversity of organisms, the processes and routes by which that diversity has been achieved and, by deduction, the possibilities for further diversification. When considering chromosomes in relation to plant evolution it must be remembered that we only have chromosomal data for about 10% of flowering plants and ferns; in most cases the information solely concerns the chromosome number, which is frequently obtained from only a few individuals of the species concerned, often just a single plant. Furthermore, species differ markedly from each other in the ease with which their chromosomes may be studied in detail. Consequently, reasonably comprehensive cytogenetical data are available for only a rather small range of the diversity of living plants and, of course, for none of those that are extinct, so that some caution is necessary when attempting to formulate general principles relating to chromosomes and evolutionary patterns.

6.1 Chromosomes and the breeding system

The importance of the breeding system in moulding the genetic architecture of populations and species is well known. Outbreeders are heterozygous and so have a wide range of genotypes, although the array of phenotypes may be reduced by dominance and epistasis. This type of breeding system provides long-term adaptability but, in the short-term, adaptive combinations of genes are broken down by recombination. Inbreeders are more homozygous and the progeny are more or less genetically identical with their parents, so that adapted genotypes can be retained intact, apart from mutation, but at the expense of long-term flexibility. Similarly, though for different reasons (Section 6.1.2), apomictic plants can produce identical progeny.

6.1.1 Amphimixis

Darlington has emphasized the importance of the *Recombination Index* (chiasma frequency per cell + gametic chromosome number) in relation to the breeding system of sexually reproducing plants and the regulation of variability. In order to preserve adaptive combinations of genes an outbreeder may be expected to restrict recombination [2] which, chromo-somally, can be achieved by a lowered frequency and increased localization of chiasmata and by a reduction in the chromosome number. The chiasmata are under genotypic control [3,4], as exemplified by the marked genetic differences in chiasma-frequency between populations in *Silene album* and *S. dioicum* [2]. In a number of groups, e.g. *Agropyron* [6], *Allium* [7], *Collinsia* [2], *Gilia* [8], *Limnanthes* [9] and *Sorghum* [10], chiasma-frequency is lower in outbreeders than in related inbreeders, while in *Allium* localized chiasmata only occurred in outbreeders. In *Secale cereale* [4], however, inbred lines had fewer chiasmata than inter-line hybrids, thus emphasizing the independent adjustment of the various components of the recombination system in response to natural selection [5].

Heterozygosity may also be reflected at the chromosome level. Interchange heterozygotes, for example, can readily be distinguished from either of the homozygotes (Section 4.4). On selfing an interchange heterozygote equal numbers of structural homozygotes and structural heterozygotes would be expected (Fig. 6.1), the latter being as genically heterozygous as the parent, the former being less so. Interchange hererozygotes of *Campanula persicifolia* [11],

Clarkia williamsonii [25] and *Secale cereale* [12], when selfed gave many more heterozygotes than structural homozygotes among the progeny, suggesting selection against genic homozygosity. However, a different interchange in *Secale* had no such selective advantage over its homozygote; clearly heterozygosity *per se* bestows no inevitable advantage — that depends on what segments and loci are heterozygous.

Some species, with all their chromosomes involved in reciprocal translocations (Section 4.4), can maintain permanent structural heterozygosity despite their autogamy by a mechanism first satisfactorily worked out in *Oenothera* [13]. Interchange heterozygotes only produce viable gametes when the chromosomes show alternate disjunction (Section 4.4). Consequently, each gamete in *Oenothera lamarkiana* (n = 7), for example, receives 7 separate but collectively inherited chromosomes which essentially constitute a single linkage group, a Renner complex. As a result of mutation and the low level of crossing-over in the differential segments the two complexes may differ appreciably in genetic content. Thus, in *O. lamarkiana* the two complexes, termed *gaudens* and *velans,* give very different hybrids when outcrossed to other species. The accumulation, in the heterozygous condition, of deleterious or lethal genes, which may resemble self-incompatibility loci [14], prevents the survival of structural homozygotes (Fig. 6.2) by zygotic lethality. Similarly,

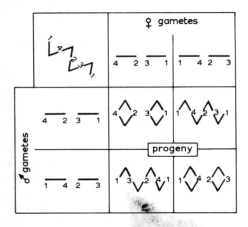

Fig. 6.1 Expected result of selfing an interchange heterozygote, showing meiotic configurations in parent and progeny.

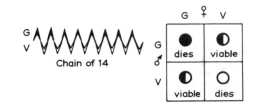

Fig. 6.2 Operation of Renner Effect in *Oenothera lamarckiana*. Alternate segregation results in gametes containing 7 chromosomes of either *gaudens* (G) or *velans* (V) set. Homozygotes are inviable.

gametic lethals can prevent the formation of homozygotes since each Renner complex can only be transmitted via either pollen or egg, thereby giving about 50% abortion of pollen and ovules, as in *Oenothera biennis, O. muricata* and *Gayophytum heterozygum* [15].

The evolution of breeding systems combining complex hybridity with inbreeding has been explained in two ways. Cleland [16] considers that heterozygosity in such cases arose by crossing between outbreeding homozygotes differing by several interchanges and that stabilization followed the subsequent evolution of autogamy and balanced lethals. Darlington [11], however, holds that complex hybridity arose gradually in response to inbreeding, as noted above. These need not be mutually exclusive alternatives: in *Isotoma petraea* [17], for example, complex hybridity was apparently initiated by hybridization between populations which gave small interchange rings that were subsequently enlarged by further interpopulational crosses to conserve hybridity as had been shown in *Campanula* [11]. Artificial hybrids between structurally homozygous populations of *Isotoma* showed marked heterosis, while those between complex heterozygotes, which are already conserving genetic heterozygosity, did not or were even negatively heterotic [18]. The need for caution in relating chromosome behaviour to breeding system is exemplified by *Gaura* [19]. The self-compatible *G. biennis* and *G. triangulata* are complex heterozygotes of the *Oenothera biennis* type, with 50% pollen abortion, and are apparently of hybrid origin, but dramatically high levels of structural heterozygosity were noted in the self-incompatible *G. calcicola, G. lindheimeri* and *G. villosa,* in which the chromosomes reflect the heterozygosity of the outbreeding system.

6.1.2 Apomixis
Apomixis includes asexual reproduction, by means of fragmentation, stolons, tubers, bulbils or pseudovivipary, and agamospermy, in which seed is produced by complete or

partial circumvention of the normal sexual processes. Facultative apomicts, such as *Rubus* and *Potentilla,* are capable of periodic sexual reproduction while obligate apomicts, such as *Alchemilla* and *Taraxacum,* are not. Agamospermy is achieved in many different ways [20] but only the major types need be mentioned here. Most commonly, female meiosis is suppressed so that the embryo is derived mitotically, either from the megaspore mother cell (diplospory), as in *Antennaria alpina* and *Arnica,* or from one or more somatic cells in the nucellus or chalaza (apospory), as in *Hieracium* subg. *Pilosella, Oxyria digyna* and *Panicum maximum,* with or without degeneration of the megaspore mother cell. Since pollination is not involved the male gametes are not exposed to selection and frequently accumulate a high level of chromosomal abnormalities.

In aneuspory, shown by *Taraxacum* and *Chondrilla,* for example, the first meiotic division occurs but the chromosomes remain in the same cell to form a restitution nucleus which, after the second division, gives a dyad of unreduced cells of which the lower ultimately forms the embryo. In these cases crossing-over during the first meiotic division can lead to reassortment of genetic material so that position effects may alter the phenotypes of the progeny, as may the loss of chromosomes following their irregular behaviour during the formation of restitution nuclei. In *Allium odorum* [21] endomitosis during meiotic prophase leads to a doubling of the chromosome number. Pairing and crossing-over apparently occur between genetically identical sister chromosomes, so that a normal meiosis produces an unreduced megaspore which gives rise to an embryo identical with the parent.

Many genera, such as *Parthenium, Potentilla* and *Rubus,* show pseudogamy, in which pollination is appare necessary for successful seed deve egg-cell is not fertilized but c clei often fuses with th embryo-sac to form e er ver, this cannot explain the role of pseudogamy in *Orchis*, which has non-endospermic seeds; clearly there is much yet to be learned about these processes.

Apomixis permits the exact reproduction of adaptive combinations and, since the sexual ancestors of most apomicts seem to be out-breeders [22], it can be interpreted as an extreme way of reducing recombination. How-ever, most apomicts are also polyploid and of hybrid origin, and the system permits their reproduction despite possible sterility and meiotic irregularity, as well as conserving heterozygosity. Indeed, in obligate apomicts the degree of heterozygosity and chromosomal irregularity can increase by mutation and structural rearrangements. Pseudogamous plants will be less prone to such changes since the pollen must remain at least partly fertile and so liable to selection, while the meiotic apparatus of facultative apomicts will be subject to strong selection during periods of sexuality and this will prevent excessive accumulation of deleterious genes. Facultative apomicts, like inbreeders that periodically outcross, can conserve adaptive gene complexes for immediate exploitation of the environment while retaining access to long-term fitness through the variability released when they revert to sexual reproduction.

6.2 Chromosomes and distribution

Once it has become established as a separate entity, a species will, if successful, compete with other species and begin to expand its area. Near its limit a species tends to be more dis-persed, so that marginal populations may be subject to a higher level of inbreeding than more 'central' populations, and this is accentuated by periodic reductions in their size as a result of environmental vicissitudes or biotic competi-tion. Consequently, mechanisms conserving or enhancing heterozygosity may be expected at the 'colonizing edge' of a species and this is amply borne out by studies of the chromosomes.

6.2.1 *Structural hetero*

In the central part of the range of *Paeonia*

Fig. 6.3 Distribution of *Paeonia californica* (- - - -) showing number of chromosome pairs not included in multiple configurations in populations studied cytologically by Walters [23].

californica, around Santa Barbara in California (Fig. 6.3), the populations are structurally homozygous and form 5 bivalents [23]. At its northern limit plants heterozygous for at least 2 or 3 interchanges are found, while at its southern limit near the Mexican border all the chromosomes are linked to form the maximum configuration of a ring-of-ten. Although this species is largely outcrossing there is clearly an advantage in making special arrangements to promote heterozygosity in its colonizing populations. A survey [25] of somatic karyo-types of the South African *Haworthia reinwardtii* var. *chalumnensis* showed a similar pattern. At the western end of its area only interchange homozygotes were found, near the point where the variety seems to have origin-ated. Further eastwards there is a gradual in-crease in the number of heterozygous inter-changes, suggesting that colonization has been accompanied, or made possible, by increasing heterozygosity.

The role of ecological factors is demonstrated by *Clarkia williamsonii* ($n = 9$), a self-compatible, normally outcrossing species of the Sierra Nevada in California [24]. Above 4000 ft it grows around meadows in the yellow pine forests, where most plants are structurally homozygous. At lower elevations it occurs in open oak-digger pine woodland where about

53

half the individuals are structural heterozygotes. The woodland has less rain than the forest and the frequent winter droughts can drastically reduce or eliminate populations of *C. williamsonii*. Experiments, although admittedly inadequate, did not suggest that heterozygotes can tolerate drought better than structural homozygotes, but they do have a selective advantage when self-pollinated (p. 51), and thus are more prevalent in the woodland where close inbreeding results from the periodic truncation of population size.

Structural heterozygosity is not necessarily more frequent at a species' margin. In *Clarkia unguiculata* [26], for example, 35% of plants sampled from natural populations were heterozygous for at least one translocation. The interchange heterozygotes occurred almost throughout the range of the species without obvious correlation with geographical or ecological factors. By examining multivalent formation in artificial interpopulational hybrids, five different arrangements of chromosome segments were found. Individuals homozygous for arrangement A were frequent and widespread, whilst homozygotes for arrangements B-E were sporadic and localized, suggesting that they have a selective advantage only when heterozygous. In *Clarkia tenella* [27] the outbreeding subsp. *araucana* is restricted to the mesic zone of southern Chile while the derived, highly autogamous subsp. *tenella* has spread widely into the drier areas of central Chile and adjacent Argentina. Populations can differ by up to 10 interchanges but all are structural homozygotes and the chromosomes do not reflect any heterozygosity. It is not known whether autogamy in this case is accompanied by higher chiasma frequency as has been found in *Limnanthes floccosa* [9], but since the species is tetraploid the conservation of heterozygosity under inbreeding is no doubt easier than if it were diploid.

6.2.2 Polyploidy

The frequent correlation between polyploidy and distribution was first noted in high latitude and high altitude floras, which have a higher than average percentage of polyploid taxa. There is nothing in the nature of polyploidy itself (Section 5.1) which would explain this apparent tolerance of the lower temperatures of such regions, as was suggested by some earlier workers [28,29]. However, under inbreeding, polyploids move more slowly towards homozygosity than diploids, thus being more effective in fixing adaptive combinations. Furthermore, since most natural polyploids are thought to be at least partly allopolyploid they will have a level of heterozygosity commensurate with their hybridity and may even display heterosis (Fig. 6.4). These features are of value to colonizers and explain the occurrence of polyploids in areas which have become available for colonization after the retreat of the Pleistocene ice sheets. Thus, in *Biscutella laevigata* [30] the diploids occur in the European lowlands and the tetraploids in the Alps, while in *Valeriana officinalis* [31] the diploids occupy the north Eurasiatic plain and the tetraploids and octoploids the Carpathians. However, there are many instances of the reverse pattern [32]. In *Anthoxanthum odoratum-alpinum,* for example, the diploids occur at higher, and the tetraploids at lower elevations in the Alps so that care is needed in identifying the colonizing edge of the species.

Like *Clarkia unguiculata* (p. 55) many species are spreading from mesic into more

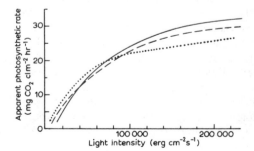

Fig. 6.4 Photosynthetic rates at different light intensities of *Mimulus nelsonii* (- - - -), *M. lewisii* (. . . .) and the artificial amphidiploid (———). (After Heisey *et al.,* [74].)

Fig. 6.5 Distribution of diploid (cross-hatched) and tetraploid (stippled) populations of *Themeda australis*, ($x = 10$). (After Hayman [33].)

xeric habitats and the colonizing role of poly-ploids in such situations is well illustrated by *Themeda australis* [33] (Fig. 6.5). This species, like many others, apparently entered Australia from the north-east and spread southwards through the more mesic tableland and coastal regions as the diploid. It subsequently moved westwards into the drier interior as the poly-ploid. The importance of considering all the characteristics of the species when interpreting cytogeographical data is emphasized by Aus-tralian *Erodium* [34], which shows a reverse pattern to *Themeda*. The diploids occupy the drier centre of Australia while the tetraploids and hexaploids spread into the better rainfall areas towards the coasts, but since *Erodium* is considered to be a dry-climate genus which probably entered Australia via an arid northern belt, the correlation between polyploidy and colonization is still maintained. The correla-tion between polyploidy and apomixis in many high latitude taxa and the breakdown of in-compatibility systems in many polyploids is also clearly pertinent to their colonizing ability.

6.2.3 Supernumerary chromosomes
The effect of supernumerary chromosomes on recombination (Section 5.4) suggests that they may be associated with colonization, particu-larly by diploids, in which supernumeraries are much more frequent than in polyploids [2]. Different frequencies of supernumerary chro-mosomes in *Festuca pratensis, Phleum phleoides* [35] and *Centaurea scabiosa* [36] have been shown to be correlated with ecological and climatic differences, but it is not clear how these are related to the dynamics of the species' distributions. However, in *Clarkia unguiculata* the populations with most super-numeraries, were those at the ecological margin of the species [37].

6.3 Chromosomes and speciation
Few evolutionists will disagree with Stebbins' [38] observation that 'five basic processes can explain evolutionary change at the level of populations and species' — mutation and chro-mosomal change, recombination, natural selection, chance fixation of genes and chromo-somal arrangements, and barriers to gene ex-change. There is still, however, disagreement about the relative importance of these pro-cesses and the evolutionary potential of dif-ferent plants will differ because of their genetic endowment, biological structure and the regions or habitats they occupy.

We have seen in the preceding chapters that the chromosomes can determine many of the processes involved in evolutionary change or can act as markers of them, but they do not do so equally in all groups or in all situations. Thus, comparable levels of reproductive isola-tion, or of spatial isolation, which are both potent factors in evolutionary differentiation, are reflected by the chromosomes in very dif-ferent ways. For instance, several temperate species and populations which occur in North and South America with comparable distribu-tion gaps across the tropics, variously showed complete interfertility and high chromosome homology, low homology and intersterility, and complete chromosome homology with low gene-exchange [39].

Polyploidy has been an important, chromo-somally documented factor in angiosperm evolution. However, polyploids are more likely to be perennials, because these can survive the initial instability and low fertility by some form of non-sexual reproduction better than annuals. Similarly, diploids are more likely than polyploids to have supernumerary chromosomes. Of the many other cases which demonstrate the restrictions upon the role of chromosomes in the evolution of different groups, a study of mitosis in the Onagraceae [40] is worth mentioning. In this family, the mitotic cycles permit the recognition of three modally distinct groups of tribes which differed in (1) the presence of chromocentres at interphase, (2) the size and uniformity of the chromosomes in the complement and, (3) the regularity of their contraction gradient during prophase. The groups comprising the tribes Onagreae and Hauyeae, in which all the genera possessing translocation systems occur, have sub equal, medium-sized, mostly metacentric chromosomes with dense proximal segments restricting chiasma formation to the distal portions of the arms. These features, present only in certain genera, permit the mechanical flexibility required if interchange systems are to be of evolutionary importance to the species.

6.3.1 Aneuploidy

A change in chromosome number, particularly at the diploid level, is usually accompanied by reduced fertility in the heterozygous condition (Section 5.3) so that, if it becomes stabilized as the homozygote, reproductive isolation and speciation can ensue. The aneuploid differences between the species of many genera demonstrate the importance of this mode of speciation and, furthermore, careful analysis can often indicate the direction of the change so adding a phylogenetic dimension to the observation.

The mechanism by which the basic number can be reduced was explained earlier (Section 5.3.2) in *Crepis*, *Haplopappus* and *Gibasis*. By similar means *Chaenactis fremontii* ($n = 5$) and *C. stevioides* ($n = 5$) independently evolved from *C. glabriuscula* ($n = 6$), with which they are morphologically very similar but occur in more xeric habitats in the S.W. United States [41]. Meiotic chromosome configurations in experimentally produced hybrids (Fig. 6.6) indicated that each of the 5-chromosome species had a compound chromosome with respect to *C. glabriuscula* and that the compound chromosomes had different structures.

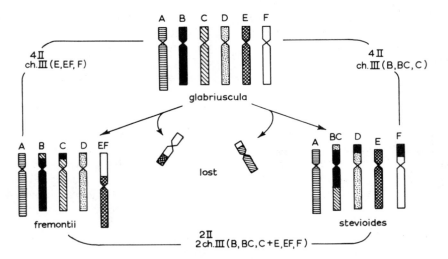

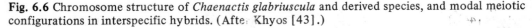

Fig. 6.6 Chromosome structure of *Chaenactis glabriuscula* and derived species, and modal meiotic configurations in interspecific hybrids. (After Khyos [43].)

The reduced chromosome number and lower chiasma frequency of the derived desert species probably preserve adaptive complexes of genes from recombination while the structural differences maintain the isolation between the three species.

Speciation accompanying an increase in basic number was demonstrated at the southern limit of the widespread and variable Californian endemic *Clarkia biloba* ($n = 8$) [44]. Here, the very local *C. lingulata* ($n = 9$) has arisen by the addition of a chromosome made up of parts of two chromosomes of the *C. biloba* genome following non-disjunction in the interchange heterozygote and subsequent selection for homozygosity among the progeny (Fig. 6.7). The species differ only in petal shape but the hybrids between them are of very low fertility. *C. lingulata* probably originated in a small population of *C. biloba* adapted to the marginal conditions and was separated from the rest of the species by the chromosomal change. *C. lingulata* has a lower reproductive capacity than its ancestor and the aneuploid change undoubtedly reduces competition with *C. biloba* [24].

Aneuploid changes can have an important role in speciation but several species, such as *Cardamine pratensis*, tolerate high levels of aneuploidy which seems to be important in the adaptability of the population. In *Claytonia virginica*, with fifty different cytotypes [43] between $2n = 12$ and $2n = c. 191$, some populations exhibit a seasonal shift in the range of chromosome numbers which can be correlated with changes in environmental conditions.

Genetical variability in the populations of most plants is maintained against a background of stable chromosome numbers but in this case the latter clearly participate in and reflect the variation.

6.3.2 Polyploidy

As noted above polyploidy has been one of the most widespread factors in the evolution of flowering plants and ferns. Not only do polyploids constitute a means for the sexual propagation of hybrids and the retention of heterozygosity but an immediate barrier to gene exchange is provided between them and their diploid ancestors. Thus, although the sterility of triploids has frequently been overestimated they reduce gene-flow between diploid and tetraploid to such an extent that differentiation can proceed in reasonable reproductive isolation. Furthermore, since diploids normally give rise to polyploids the direction of evolution can be observed, although in view of polyhaploidy this cannot be automatically assumed [44]; in *Dichanthium*, for example, diploids have been shown to arise from tetraploids [45].

Polyploids, even allopolyploids, do not arise fully fertile from their diploid forbears (Section 5.1). Their association with vegetative reproduction and the perennial habit points to their need to survive a period of partial sterility before they become stabilized. This adjustment is also evident with regard to the bulk of chromosomal material which the cells can accommodate. With unrestricted increase in ploidy level the cells would eventually have little room for anything but chromosomes and the irregular

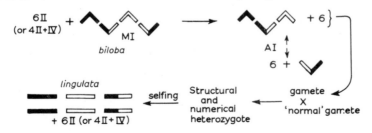

Fig. 6.7 Probable origin of *Clarkia lingulata* ($n = 9$) from *C. biloba* ($n = 8$). (After Lewis and Roberts [44].)

mitotic and meiotic behaviour of higher polyploids is a reflection of the strains they put on the cell mechanisms. There are several observations, however, which indicate one way in which the increased chromosome number can be tolerated. In Section 3.2 it was seen that cells with twice the chromosome number of related cells had twice the DNA content, and in *Lotus* [46], *Avena* [47], and *Triticum* [48], for example, diploids and derived tetraploids and hexaploids show a similar simple relationship. However, in *Leucanthemum* [49] the DNA values of $2x$, $4x$ and $8x$ species fitted a ratio of 1:0:1.98:2.5 and not 1:2:4 as expected from the chromosome numbers, while in *Betula* there was a direct correlation between chromosome number and DNA content in the 28-, 42-, 56- and 70-chromosome plants, but the 84-chromosome plants had a DNA value expected of a plant with 61 chromosomes [50]. Apparently, cells can tolerate chromosome changes to some degree but beyond that compensation in the total DNA value seems necessary (Section 6.4).

The experimental induction of allopolyploids, including the resynthesis of such naturally occurring species as *Galeopsis tetrahit* (Section 5.1), has provided convincing proof of their evolutionary role and mode of origin. Despite this, much more information is needed to chart the processes involved in the origin of most natural polyploids, for which only the chromosome number is known. Even apparently clear cases can be deceptive. *Poa annua* ($2n = 28$) is morphologically intermediate between *P. infirma* ($2n = 14$) and *P. supina* ($2n = 14$) and, since its hybrids with each of them showed 7 bivalents and 7 univalents at meiosis, it was considered to be an allotetraploid [51]. However, the discovery of diploid populations of *P. annua* with regular bivalent formation [54], as well as observations on somatic karyotypes [53], suggested a contrary explanation and it is still not clear whether this species is allopolyploid or autopolyploid.

Where homoeologous pairing and other factors affecting syndesis do not cloud the picture unduly, analysis of meiotic configurations in artificial hybrids can be of great value in unravelling the relationships between species involved in a polyploid complex (Fig. 6.8). Such genome analysis demonstrates the reticulate pattern of evolution open to so many polyploid plants.

Between the situation just illustrated, in which diploids and polyploids can be readily recognized as specifically distinct, and that found in *Arenaria ciliata* and *Campanula rotundifolia,* in which they are morphologically almost indistinguishable, groups of diploids and related polyploids can show confusing patterns of variation. These can often be interpreted in terms of differentiation-

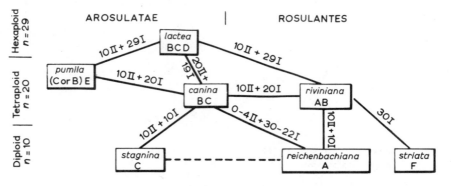

Fig. 6.8 Genomic constitutions of species in *Viola* subsect. *Rostratae* and chromosome-pairing in hybrids. Unsuccessful crosses shown by broken lines. (Partly after Moore and Harvey, [75].)

hybridization cycles [54] which give rise to so-called polyploid pillar complexes. At the diploid level, differentiation and the establishment of genic and chromosomal barriers to crossing gives rise to a series of discontinuous races and species. Polyploid derivatives can blur these differences, partly because of the hybridity of any allopolyploids formed and partly because hybridization is easier at the polyploid level. Consequently, in groups such as the *Achillea millefolium* complex [54], *Gayophytum diffusum* [15] and *Galium graecum-canum* [55], the morphologically distinguishable diploids give rise to polyploids which resemble their diploid progenitors and also hybridize with each other to obscure the original differences (Fig. 6.9). Furthermore, a period of stabilization can lead to differentiation among the polyploids which can then give rise to higher polyploids and a yet further cycle

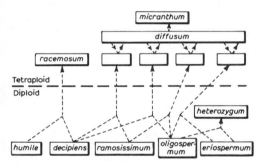

Fig. 6.9 Polyploid pillar complex in *Gayophytum*. (After Lewis and Szweykowski [15].)

of hybridization. Although posing certain taxonomic problems, differentiation-hybridization cycles are clearly a potent form of reticulate evolution which can be interpreted only with adequate cytogenetical information.

6.4 Chromosomes and major evolutionary trends

The most complete cytogenetical data can only be obtained when the behaviour and homology of the chromosoms can be observed in hybrids.

Consequently, such data have been used most fully in attempts to understand the evolution of relatively closely allied groups of species — the so-called micro-evolution considered in the previous sections. Because of this, other forms of data are generally used to interpret broad evolutionary trends in plants and the chromosomes are frequently of little help.

It is clear from its frequency in angiosperms that polyploidy has been important in their evolution, as it has in pteridophytes, while the gain and loss of centromeres has given rise to the aneuploid series constituting the various basic numbers in many families and genera. The starting point (or points) for these changes, the much-debated original basic number of the angiosperms, has been considered [57] to be in the range $x = 7$-9, which is most frequent in modern representatives. Indeed, studies of various families in the Magnoliales, which are often considered to show the highest proportion of primitive morphological and anatomical characters, have led to the more precise suggestion [58,59] that the earliest angiosperms had $x = 7$. Furthermore, the occurrence of symmetrical karyotypes, composed of relatively large submetacentric chromosomes, in some of these families, such as the Himantandraceae [60], has been taken in conjunction with data from the Compositae-Cichorieae [61], Dipsacaceae [62] and Ranunculaceae-Helleboreae [63] to suggest that during angiosperm evolution karyotypes have become less symmetrical and the chromosomes smaller. However, in the Podocarpaceae [64], for example, karyotypes appear to have evolved from asymmetry to symmetry, while the often considerable variation in chromosome size between families, genera or even species has not yet been satisfactorily explained and does not suggest any simple evolutionary trend.

The size of the chromosomes will partly depend upon the amount of DNA they contain, as has been demonstrated in *Lathyrus, Anemone, Vicia* and *Lolium* [65, 67], and might be expected to be correlated with the genetical complexity of the organism; certainly cells of higher plants and animals have much more DNA than do those of primitive phyla such as bacteria. However, the differences between higher plants, particularly related species of the same chromosome number, cannot readily be explained in this way. There is some suggestion [67] that such differences in DNA content are due to lateral replication of chromatids, i.e. polynemy, but there is more convincing evidence that they result from longitudinal duplication [66, 68]. Much of this duplication involves the repetitive DNA of heterochromatin (Section 2.3; [69, 70]), but euchromatin cannot be excluded. Evidence as to whether the amount of DNA increases or decreases during evolution can be obtained by comparing less and more derived taxa in relation to reasonably well established evolutionary trends, such as that from outbreeding to inbreeding or from the perennial to the annual habit. No single relationship has been found. In *Lolium* inbreeders have more DNA than outbreeders, but the reverse is true in *Lathyrus* [70], while in Compositae-Anthemideae annuals can have more or less DNA than related perennials [69]. Very high DNA amounts, however, seem to be strongly correlated with the perennial habit, while species with lower DNA values are either annual or perennial depending upon the genotype [71]. To what extent these trends are due to genetic complexity or to the loss of redundant genetic material is not known. Certainly the length of the cell-cycle, which is affected by the amount of DNA (Section 3.1), and the chiasma-frequency, which is affected by the balance between eu- and heterochromatin (Section 3.3), must have evolutionary importance. Furthermore, the relation between the proteins of chromatin and changes in chromosome-size are not understood but cannot be ignored.

In conclusion, it is perhaps worth re-emphasizing that, despite occasional haploidization, the change from diploidy to polyploidy provides chromosomal evidence of an evolutionary trend of great importance in plants. With adequate

cytogenetical information, structural rearrangements and gain or loss of centromeres can be similarly interpreted, but their occurrence and the direction of change varies from group to group, as is clearly true of those features of the chromosomes just discussed. It has been repeatedly stressed that different groups of plants exhibit different evolutionary strategies, which may or may not be reflected in the chromosome so that attempts to discern inevitable trends in karyotype evolution are at best premature [73], given the paucity of chromosome data from so many groups and the continuing uncertainty about the origin and diversification of the angiosperms. The cytogenetics of plants, as of animals, provides a unifying foundation for understanding how genetic information is transmitted, within what parameters there is scope for evolutionary change and, to some extent, how that change has come about. However, much more work is necessary to properly chart the relationship between the evolution of the diversity of plants and their chromosomes which bear the genetic determinants.

References

[1] Darlington, C.D. (1937), *Recent Advances in Cytology,* edit. 2, Churchill, London.
[2] Lewis, K.R. and John, B. (1963), *Chromosome Marker,* Churchill, London.
[3] Darlington, C.D. (1932), *Amer. Nat.,* **96**, 25-51.
[4] Rees, H. (1961), *Bot. Rev.,* **27**, 288-318.
[5] Lawrence, C.W. (1963), *Heredity,* **18**, 135-147.
[6] Stebbins, G.L., Valencia, J.I. and Valencia, R.M. (1946), *Amer. J. Bot.,* **33**, 579-586.
[7] Ved Brat, S. (1965), *Heredity,* **20**, 325-339.
[8] Grant, V. (1954), *El Aliso,* **3**, 1-49.
[9] Arroyo, M.T.K. (1973), *Evolution,* **27**, 679-688.
[10] Garber, E.D. (1960), *Cytologia,* **25**, 233-243.

(1950), *Heredity,* **4**, 217-248.

[12] Thompson, J.B. and Rees, H. (1955), *Nature (Lond.),* **177**, 385.

[13] Renner, O. (1921), *Zeits. Bot.,* **13**, 609-621.

[14] Steiner, E. (1961), *Genetics,* **46**, 301-315.

[15] Lewis, H. and Szweykowski, J. (1964), *Brittonia,* **16**, 343-391.

[16] Cleland, R.E. (1960), *Proc. Indiana Acad. Sci.,* **69**, 51-64.

[17] James, S.H. (1970), *Heredity,* **25**, 53-77.

[18] Beltran, I.C. and James, S.H. (1974), *Austral. J. Bot.,* **22**, 251-264.

[19] Raven, P.H. and Gregory, D.P. (1972), *Brittonia,* **24**, 71-86.

[20] Battaglia, E. (1963), In: Maheshwari, P. (ed.), *Recent Advances in the Embryology of Angiosperms,* Delhi, 221-264.

[21] Häkansson, A. and Levan, A. (1957), *Hereditas (Lund),* **43**, 179-200.

[22] Stebbins, G.L. (1951), *Variation and Evolution in Plants,* Columbia Univ. Press, New York.

[23] Walters, J.L. (1942), *Amer. J. Bot.,* **29**, 270-275.

[24] Lewis, H. (1969), *BioScience,* **19**, 223-227.

[25] Brandham, P.E. (1974), *Chromosoma (Berl.),* **47**, 85-108.

[26] Mooring, J.S. (1958), *Amer. J. Bot.,* **45**, 233-242.

[27] Moore, D.M. and Lewis, H. (1966), *Heredity,* **21** 37-56.

[28] Tischler, G. (1935), *Bot. Jahrb.,* **67**, 1-36.

[29] Müntzing, A. (1936), *Hereditas (Lund),* **21**, 263-378.

[30] Manton, I. (1934), *Zeits. Ind. Abst. Vererbungsl.,* **67**, 41-57.

[31] Skalinska, M. (1950), *Acta Soc. Bot. Pol.,* **20**, 45-68.

[32] Favarger, C. (1967), *Biol. Rev.,* **42**, 163-206.

[33] Hayman, D.L. (1960), *Austral. J. Bot.,* **8**, 58-68.

[34] Carolin, R.C. (1958), *Proc. Linn. Soc. N.S.W.,* **83**, 92-100.

[35] Bosemark, N.O. (1956), *Hereditas (Lund),* **42**, 189-210, 443-466.

[36] Fröst, S. (1958), *Hereditas (Lund),* **44**, 75-111.

[37] Mooring, J.S. (1960), *Amer. J. Bot.,* **47**, 847-854.

[38] Stebbins, G.L. (1974), *Flowering Plants, Evolution above the species level,* Arnold, London.

[39] Moore, D.M. and Raven, P.H. (1970), *Evolution,* **24**, 816-823.

[40] Kurabayashi, M., Lewis, H. and Raven, P.H. (1962), *Amer. J. Bot.,* **49**, 1003-1026.

[41] Kyhos, D.W. (1965), *Evolution,* **19**, 26-43.

[42] Lewis, H. and Roberts, M.R. (1956), *Evolution,* **10**, 126-138.

[43] Lewis, W.H. (1970), *Science,* **168**, 1115-1116.

[44] Raven, P.H. and Thompson, H.J. (1964), *Amer. Nat.,* **98**, 251-252.

[45] De Wet, J.M.J. (1968), *Evolution,* **22**, 394-397.

[46] Cheng, R.I-J. and Grant, W.F. (1973), *Can. J. Genet. Cytol.,* **15**, 101-115.

[47] Yang, D.P. and Dodson, E.O. (1970), *Chromosoma (Berl.),* **31**, 309-320.

[48] Rees, H. and Walters, M.R. (1965), *Heredity,* **20**, 73-82.

[49] Probst, F. (1971), *Chromosoma (Berl.),* **36**, 322-328.

[50] Taper, J.L. and Grant, W.F. (1973), *Caryologia,* **26**, 263-273.

[51] Tutin, T.G. (1957), *Watsonia,* **4**, 1-10.

[52] Ellis, W.M., Calder, D.M. and Lee, B.T.O. (1970), *Experientia,* **26**, 1156.

[53] Koshy, T.K. (1968), *Can. J. Genet. Cytol.,* **10**, 112-118.

[54] Ehrendorfer, F. (1959), *Cold Spr. Harb. Symp. Quant. Biol.,* **24**, 141-152.

[55] Ehrendorfer, F. (1958), *Osterr. Bot. Zeits.,* **105**, 229-279.

[56] Davis, P.H. and Heywood, V.H. (1963), *Principles of Flowering Plant Taxonomy,*

Oliver and Boyd, Edinburgh.

[57] Grant, V. (1963), *The Origin of Adaptations,* Columbia, New York.

[58] Raven, P.H. and Khyos, D.W. (1965), *Evolution,* **19**, 244-248.

[59] Ehrendorfer, F., Krendl, F., Habeler, E. and Sauer, W. (1968), *Taxon,* **17**, 337-353.

[60] Sauer, W. and Ehrendorfer, F. (1970), *Osterr. Bot. Zeits.,* **118**, 38-54.

[61] Stebbins, G.L. (1958), *Cold Spr. Harb. Symp. Quant. Biol.,* **23**, 365-378.

[62] Ehrendorfer, F. (1965), *Genetics Today,* **2**, 399-407.

[63] Lewitsky, G.A. (1931), *Bull. Appl. Bot. Genet. Pl. Breed.,* **27**, 220-240.

[64] Hair, J.B. and Beuzenberg, E.J. (1958), *Nature (Lond.),* **181**, 1584-1586.

[65] Rees, H. and Hazarika, M.H. (1969), *Chromosomes Today,* **2**, 158-165.

[66] Rees, H. and Jones, R.N. (1972), *Internat. Rev. Cytol.,* **32**, 53-92.

[67] Rothfels, K., Sexsmith, E., Heimbuger, M. and Krause, M.O. (1966), *Chromosoma (Berl.),* **20**, 54-74.

[68] Stucky, J. and Jackson, R.C. (1975), *Amer. J. Bot.,* **62**, 509-518.

[69] Nagl, W. and Ehrendorfer, F. (1974), *Plant Syst. Evol.,* **123**, 35-54.

[70] Miksche, J.P. and Hotta, Y. (1973), *Chromosoma (Berl.),* **41**, 29-36.

[71] Smith, J.B. and Bennett, M.D. (1975), *Heredity,* **35**, 231-239.

[72] Stebbins, G.L. (1966), *Science,* **152**, 1463-1469.

[73] Jones, K. (1970), *Taxon,* **19**, 172-179.

[74] Hiesey, W.M., Nobs, M.A. and Björkman, O. (1969), *Carn. Inst. Yrbk.* **67**, 489-491.

[75] Moore, D.M. and Harvey, M.J. (1961), *New Phytol.,* **60**, 85-95.

Suggestions for further reading

Throughout this book the references listed after each chapter document the statements made and the data utilized. An outline text such as this can only provide a framework of criteria for judging the fuller information continually being provided in original papers. As far as possible the student should make a habit of scanning the articles in such journals as *Chromosoma, Heredity* and others whose titles appear frequently in the references. Only in this way is it possible to keep abreast of the changing trends in cytogenetics and to gain some insight of the triumphs and frustrations of such research.

[1] Darlington, C.D. and La Cour, L.F. (1962), *The Handling of Chromosomes,* 3rd edition, Allen and Unwin, London.

One of the best accounts of the various techniques available for observing the chromosomes. It is a most useful *'vade mecum'* which should be at the side of everyone who works with chromosomes.

[2] MacLeish, J. and Snoad, B. (1958), *Looking at Chromosomes,* MacMillan, London.

A description of mitosis, meiosis and the development of the pollen grain and embryo-sac in *Lilium regale,* illustrated by a series of excellent photographs.

[3] Lewis, K.R. and John, B. (1973), *The Matter of Mendelian Heredity,* 2nd edition, Longmans, London.

In their distinctively lucid and readable style these authors present a clear account of the behaviour of genes and chromosomes in relation to breeding systems and selection. There is also a useful chapter on cytological techniques.

[4] Whitehouse, H.L.K. (1974), *Towards an Understanding of the Mechanism of Heredity,* 3rd edition, Arnold, London.

A well-documented and detailed treatment. The first eight chapters are especially pertinent to this book.

[5] John, B. and Lewis, K.R. (1965), *The Meiotic System,* Springer-Verlag, Wien.

[6] John, B. and Lewis, K.R. (1968), *The Chromosome Complement,* Springer-Verlag, Wien.

[7] John, B. and Lewis, K.R. (1969), *The Chromosome Cycle,* Springer-Verlag, Wien.

This trilogy, which appeared as *Protoplasmatologia* VI:F1, VI:A and VI:B, respectively, provides an essential source of ideas and data on cytogenetics. Well illustrated with photographs and diagrams.

[8] Rieger, R., Michaelis, A. and Green, M. M. (1968), *A Glossary of Genetics and Cytogenetics,* Allen and Unwin, London.

A scholarly volume which should be consulted by anyone concerned with cytogenetics who wishes to avoid misuse of words and misconception of meanings.

[9] Darlington, C.D. (1963), *Chromosome Botany and the Origins of Cultivated Plants,* 2nd edition, Allen and Unwin, London.

As one of the most vigorous proponents of cytogenetics for over forty years, the author gives here a sometimes contentious but always informative and stimulating view of the role of chromosomes in the biology and evolution of plants.

[10] Stebbins, G.L. (1971), *Chromosomal Evolution in Higher Plants,* Arnold, London.

A very readable account by one of the foremost students of plant evolution. His *Variation and Evolution in Plants* (1951), Columbia, New York, of which this book is a direct descendant, is still one of the most masterly syntheses in the field.

Index